CAMBRIDGE STUDIES IN MODERN BIOLOGY: 2

THE ECOLOGY OF MARINE SEDIMENTS

JOHN S. GRAY
Institute for Marine Biology and Limnology, University of Oslo, Norway

THE ECOLOGY OF MARINE SEDIMENTS

An introduction to the structure and function of benthic communities

CAMBRIDGE UNIVERSITY PRESS
Cambridge
London · New York · New Rochelle
Melbourne · Sydney

Published by the Press Syndicate of the University of Cambridge
The Pitt Building, Trumpington Street, Cambridge CB2 1RP
32 East 57th Street, New York, NY 10022, USA
296 Beaconsfield Parade, Middle Park, Melbourne 3206, Australia

First published 1981

Photoset and printed in Malta by Interprint Limited.

British Library Cataloguing in Publication Data

Gray, John S
The ecology of marine sediments. – (Cambridge
studies in modern biology; 2).
1. Marine ecology
2. Benthos
I. Title II. Series
574.5′2636 QH541.5.S3 80-40869

ISBN 0 521 23553 7 hard covers
ISBN 0 521 28027 3 paperback

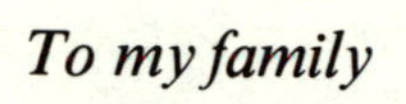
To my family

CONTENTS

PREFACE

It is a well-known fact that the sea covers three-quarters of the face of the earth. If a third dimension is added – the sea-bed, with its canyons and slopes – the sea comprises an even greater proportion of the surface of the earth. Most of the sea-bed consists of sediments and only a relatively small proportion is rocky or constructed of coral. Yet, the emphasis of ecological research has been firmly on the fauna and flora of hard substrata. In a way this is not surprising, since on a rocky intertidal shore one can see the fauna and flora and can count them directly and non-destructively; and most species are described. By contrast the fauna of sediments lie buried usually, must be sampled blind and destructively, and, at most, statistical estimates of abundance are obtained, typically with very wide confidence limits. Also, many taxonomic problems still remain, particularly with the small microscopic species constituting the microfauna and meiofauna.

In recent years great strides have been made in our understanding of rocky-shore ecology by the application of techniques and theory arising largely in terrestrial ecology. Indeed the field of research has progressed so far that now terrestrial ecologists are turning to the rocky-shore ecologists for new insights. In particular, predation theory, stemming from the works of Connell and Paine and their co-workers on the US west coast, can be singled out to illustrate this trend.

Sediment ecologists have made some important contributions to general ecological theory, such as the debate on the factors leading to high diversity in the tropics and deep sea that was stimulated by Howard Sanders (see Chapter 6). However, in the main, sediment ecologists have ignored theoretical aspects of ecology. This surprising and lamentable fact provoked Mills to claim that 'Despite more than a century of intensive work on the collection and classification of shallow water benthic animals, much of benthic ecology seems a rather shabby and intellectually suspect branch of biological oceanography. Its methods are, for the most part, those of the nineteenth century; its results, too often, are of interest only to other students of the benthos; and its importance to other branches of biological oceanography has, in my opinion, been proportionately rather small, in spite of one origin of this discipline as a branch of fisheries research.' Unfortunately, I have to agree with Mills!

This book is written to try to redress the balance by bringing to the attention of young research workers some approaches to sediment ecology possibly different from those traditionally used. It is not intended as a comprehensive review of the ecology of benthic communities, but more as an introduction to the subject. Where possible, attention is focused on new and promising research fields, such as the experimental manipulation of communities in which the importance of competition and predation in structuring the community has only recently been appreciated (see Chapter 10). These techniques stem directly from the rocky-shore work of Connell and Paine in America. In general, American workers already follow the approaches to be outlined; the book is aimed, therefore, at the European student, and as a consequence I have taken a European bias in the examples used.

ACKNOWLEDGEMENTS

I would like to thank John Mattocks, who by his enthusiasm and teaching skills started me on my career as a marine biologist, Professor Dennis Crisp, FRS, who inspired and guided my early endeavours as a research worker, and my colleagues and students, in particular Tom Pearson, Ragnar Elmgren, Knut Roed, John-Arthur Berge and Hartwig Christie, who have helped greatly by their constructive criticisms of the ideas presented here.

My thanks are also due to all those who gave permission to reproduce their original figures. I thank Richard Barnes for his timely criticisms and help in the preparation of this book. Finally, I am indebted to May Haugstad who typed my manuscript and to Jane Farrell of Cambridge University Press who corrected my grammar and pointed out many cases of lack of clarity in my writing. Any errors remaining are entirely my own.

1

The fauna of sediments

When walking down an intertidal sandy beach most people are aware that life exists within the sand, since there are often the tell-tale marks of holes, pits and mounds caused by the activities of the inhabitants. If the beach has a gentle slope with fine sand and standing pools then the evidence of this activity can be highly dramatic, with a mass of changing contours caused by various organisms. Fig. 1.1 illustrates a typical intertidal beach in northern Europe where the principal agent causing the topographic variety is the lug-worm, *Arenicola marina*. Anglers can often be seen scout-

Fig. 1.1. Pits and depressions in an intertidal mudflat caused by activities of *Arenicola marina*. (Photograph by R. S. K. Barnes.)

ing the sand with keen eyes, looking for the two adjacent holes produced by the siphons of the razor shell, *Ensis*, which they use for bait, or thriftily combing the beach for the shallow depressions made by the cockle, *Cerastoderma edule*. These, then, are the common and easily observed marks of the beach inhabitants. A closer examination of almost any beach will show minute holes produced by a profusion of amphipod crustaceans and small polychaete worms, and often rings of sand that have been formed by an animal, usually a polychaete, lying with its head downwards and its tail sticking up, engulfing sediment and defecating at the surface. With snorkel and goggles the diver can ascertain that the patterns continue below the tide marks, and in fact persist right into the deep sea.

The larger animals that leave these tell-tale patterns are usually called the macrofauna, and can be separated from the sand by sieving the sediment through a fine screen which does not allow the animals to pass through. If the sand is first separated into different depth layers then the distribution of the animals can be mapped. From this a reconstruction of the mode of life of the species can be obtained. Fig. 1.2 shows just how complex the distribution patterns can be (though admittedly the locality shown here, a subtropical beach in Georgia, is a very rich one). The commonest animals are the polychaete worms, followed by the bivalve molluscs, amphipod and decapod crustaceans, burrowing holothuroid echinoderms, and an occasional burrowing anemone.

Yet the macrofauna is only a part of the fauna of sediments. Living between the sand grains, or on muddy beaches within the mud, is a whole variety of small animals that will pass through the meshes of the screen. These small animals are called the interstitial fauna, since they live in interstices between sand grains, but the general names for all the small animals are the meiofauna and microfauna. The prefix 'meio' comes from the Greek word *meios*, meaning intermediate, and thus the meiofauna is intermediate in size between the macrofauna and microfauna. The exact definitions of macro-, meio- and microfauna have been the subject of continuing controversy over many years. Some workers use a 1 mm screen to separate out the macrofauna, whereas others use a 0.5 mm screen. Naturally, the smaller screen collects more animals. Just how big an effect this can have depends often on the season of sampling, since in periods of high larval recruitment the finer screen will collect many more juveniles. As an illustration of the sorts of results obtained, Table 1.1 shows a sample taken in California. Nematode worms cannot be sampled with a 0.5 mm screen and nor can many of the crustaceans. These two groups are what are usually referred to as meiofauna. The lower limit of the meiofaunal size is set by use of a 0.062 mm screen (biologists follow geologists in using a decreasing geometric scale of screens: 1 mm, 0.5 mm, 0.25 mm, 0.125 mm, 0.062 mm etc.). The meiofauna usually consists of nematodes, harpacticoid copepods, turbellarians, and a phylum unique to the meiofauna, the Gastrotricha. Since the meiofauna is defined by size

Fig. 1.2. Location of most important benthic animals, burrow and tubes in a shallow-shelf environment in Georgia, USA. A.a., *Abra aequalis*; C.a., *Capitomastus* cf. *aciculatus*; C.b., *Calianassa biformis*; D.c., *Diopatra cuprea*; G.a., *Glycera americana*; H.e., *Hemipholis elongata*; M.s., *Magelona* sp.; M.t., *Mesochaetopterus taylori*; N.l., *Notomastus latericeus*; O.f., *Owenia fusiformis*; O.s., *Oxyurosthylis smithi*; P.c., *Pinnixa chaetopterana*; P.g., *Pectinaria gouldi*; S.b., *Spiophanes bombyx*; S.o., *Spiochaetopterus oculatus*; T.t., *Tellina* cf. *texana*. (After Dörjes & Howard, 1975.)

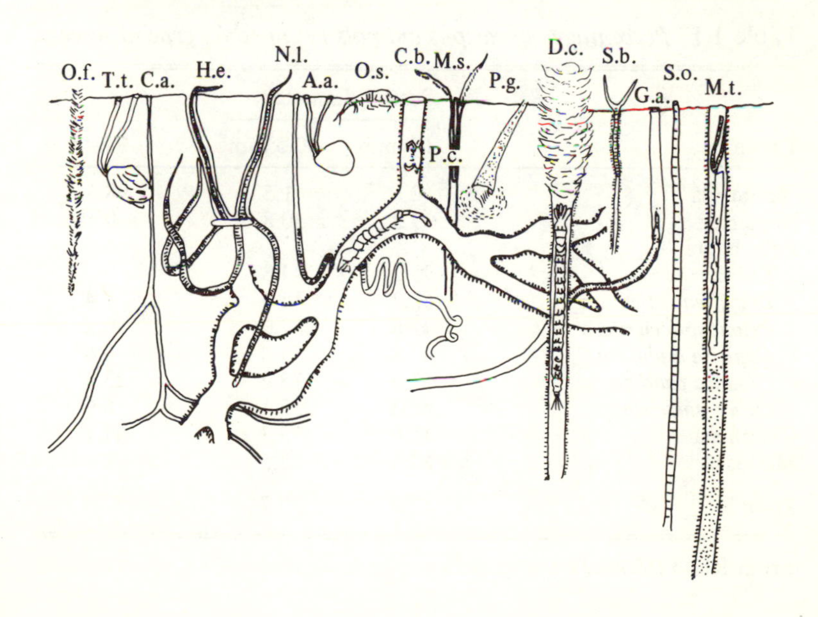

of screen used, the animals included in it may be juvenile members of the macrofauna which for a time are within the meiofaunal size range. Such animals are called temporary meiofauna and include in particular the larvae of polychaetes and bivalves, although most phyla are represented. The permanent members of the meiofauna (nematodes and harpacticoids, etc.) are those that always remain within the meiofaunal size range. In fact, there are representatives of almost all marine invertebrate phyla in the permanent meiofauna: small sponges, ascidians, gastropod molluscs, and even a walking bryozoan, *Monobryozoon ambulans*, can be found. Swedmark (1964) has reviewed the structural adaptations, and McIntyre (1969) and Fenchel (1978) the ecology of meiofauna.

The last category of sediment fauna is the microfauna, defined as that passing through a 0.062 mm screen. In practice the microfauna is not extracted by screening, and elutriation methods are usually used. The microfauna is almost exclusively ciliate protozoans.

How abundant are the different size categories in a typical sandy beach? Comparative figures are hard to find since most workers concentrate on the macrofauna alone, some (an increasing

Table 1.1. *Percentages of various animals retained on graded screens*

	Screen diameter		
Taxon	1.0 mm	0.5 mm	Residue
Nematoda	0	1.5	98.5
Nemertea	69.2	30.8	0
Polychaeta			
Lumbrinereis	95.2	4.8	0
Dorvillea articulata	62.2	34.4	3.4
Prionospio cirrifera	42.8	57.0	0.2
Capitita ambiseta	45.8	53.6	0.6
Cossura candida	1.4	75.2	23.4
Other polychaetes	58.3	35.1	6.6
Crustacea	17.6	35.3	47.1
Mollusca	87.5	12.5	0
Total	37.0	30.7	32.3

From Reish (1959).

number) are interested in the meiofauna, and only very few study the microfauna. Fig. 1.3 shows data from a typical intertidal beach where the smallest animals, the microfauna, dominate numerically, but the macrofauna dominates in terms of biomass. The actual ratios found depend on the sediment type, with, for example, microfauna being very common in fine sand but scarcer in mud, where macrofauna and meiofauna dominate. As to the species found, a typical boreal sand beach may contain 20–30 macrofaunal species and 200–300 meiofaunal species; no one has yet estimated the number of microfaunal species. The taxonomic problems of working with the meiofauna and microfauna are large, and new species are still being described regularly in these groups. Emphasis has been given here to data from intertidal beaches, not because the fauna is especially rich there, but because such beaches are better studied than subtidal areas, where boats and grabs are needed to sample the depths below 100 m. Subtidally the number of species of macrofauna in a typical sample is much higher than in the corresponding sample size taken intertidally, and in an average sampling programme of say 8–10 grab samples, each about 0.1 m^2, there will

Fig. 1.3. Abundance and biomass ranges of macro-, meio- and microfauna from sublittoral sandy sediments. (From Fenchel, 1978.)

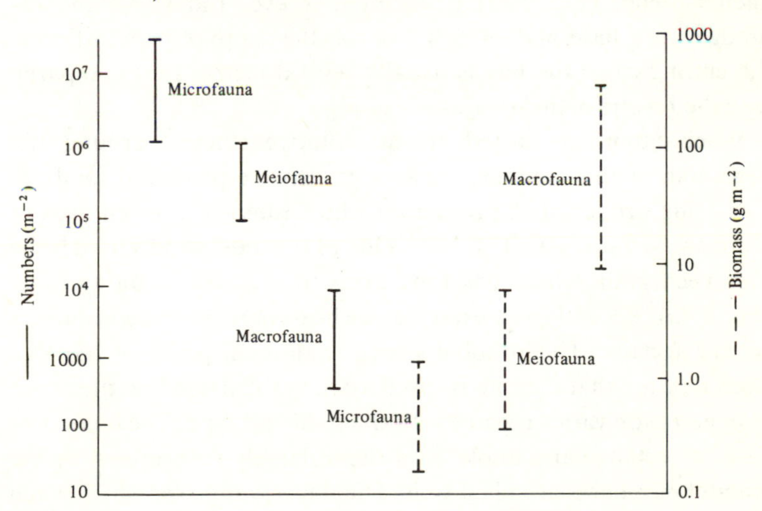

be around 150 species. Few data are available on numbers of subtidal meiofaunal and microfaunal species.

1.1 Sampling the fauna

An excellent account of the statistics of sampling the benthos is available (Elliot, 1971) and I therefore do not propose to consider such problems, but rather to describe some recent developments with important consequences for sampling the fauna.

The basic method for sampling intertidal beaches is easy: sediment is removed and the animals are extracted by screening them from the sediment. Quantitative samples are taken by removing a given sediment area, usually 0.1 m^2 for macrofauna and around 38.5 cm^2 (a 7 cm diameter coring tube) for meiofauna. Subtidally sampling is blind and done by means of gear suspended from vessels, or by scuba diving if possible. In the latter case the same methods as for the intertidal can be used.

When investigations of sediment-living fauna first began, taxonomy was naturally the central interest and so large samples giving the widest possible variety of species were preferred. The triangular naturalists' dredge was and still is one of the most effective tools for obtaining qualitative samples. In studies of the fauna of the deep sea, where obtaining one haul can take a whole day due to the long lowering and hauling time, a modified form of this dredge, the anchor-dredge (Fig. 1.4*b*), is widely used even today. An anchor-dredge has a base plate which controls the depth of penetration of the cutting edge; the bag is usually several metres long compared with the 1 metre of the naturalists' dredge.

When attention turned to quantitative studies, around the beginning of this century, various grabs were produced (and are still being produced in profusion) which sampled a given area of sediment surface, usually 0.1 m^2. One of the most widely used is the Van Veen grab, which has long arms to increase its biting force (Fig. 1.4*a*). Recently a new device, the box-corer, has received widespread acclaim. The problems with traditional grabs of the Van Veen type are that the bite is often asymmetrical and that the depth of penetration varies greatly with the sediment type. The box-corer takes a rectangular sample to a depth largely determined by the amount of weights attached to it. There is a swing arm which when

Fig. 1.4. (*a*) Van Veen grab for quantitative sampling.
(*b*) Anchor-dredge for large semi-quantitative sampling.
(*c*) Reineck box-corer for quantitative samples that leave the sediment relatively undisturbed.

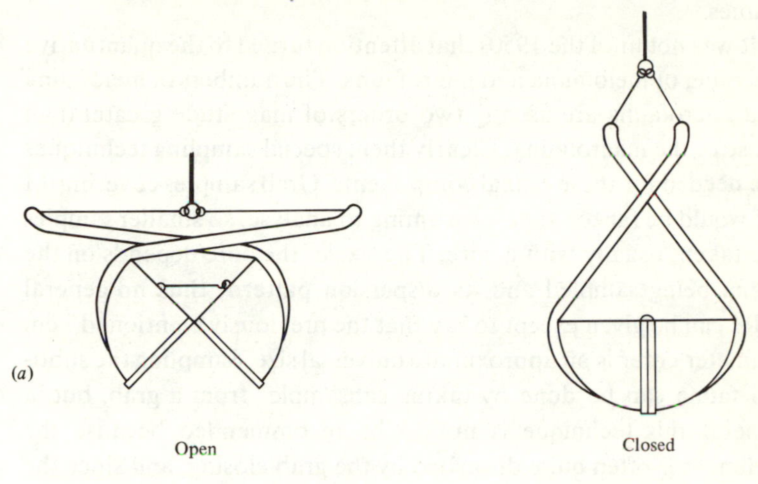

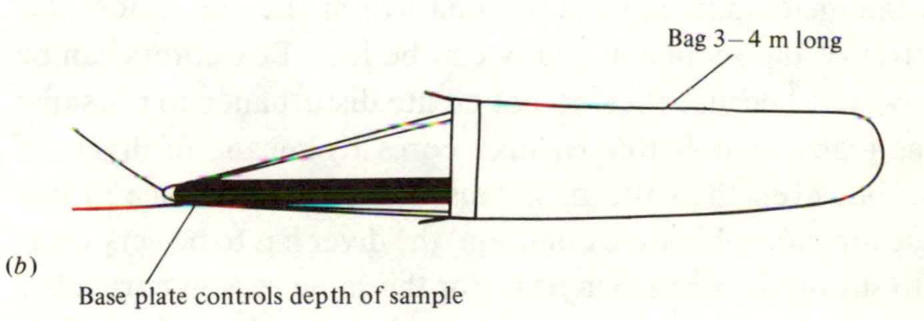

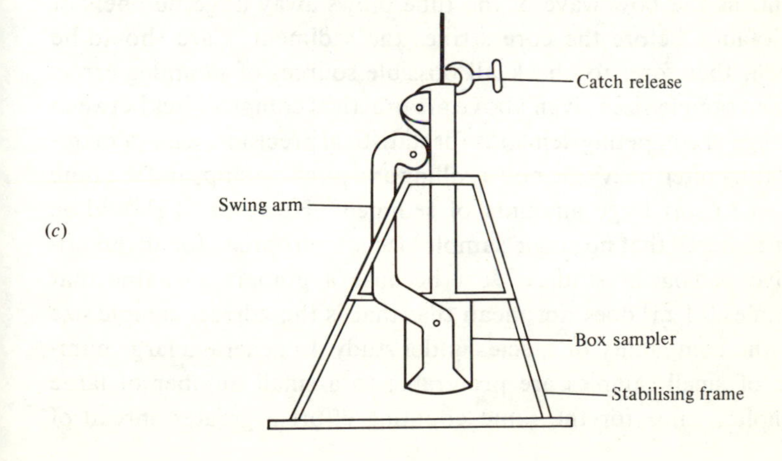

the corer is triggered closes the sampler underneath and prevents the sample sliding out (Fig. 1.4*c*). Box-corers are now widely used in both deep sea and shallow water for biological and geological studies.

It was not until the 1950s that attention turned to the quantitative sampling of meiofauna and microfauna. The numbers of meiofauna and microfauna are usually two orders of magnitude greater than those of the macrofauna. Clearly then, special sampling techniques are needed for these faunal components. Grab samples covering 0.1 m^2 would be far too time-consuming to analyse, so smaller samples are taken, usually with a core. The size of the core depends on the fauna being sampled and its dispersion pattern; thus no general rules can be given except to say that the previously mentioned 7 cm diameter corer is an approximate universal size. Sampling the subtidal fauna can be done by taking subsamples from a grab, but in general this technique is not to be recommended because the sediment is often quite disturbed by the grab closing, and since the bulk of the meiofauna and microfauna live in the uppermost few centimetres of the sediment, many can be lost. Box-corers can be used, however, because they do not create disturbance to the same extent as grabs. It is better to take cores by means of divers, if possible, but even then the problems can be severe since in fine muddy sediments (which are common) the diver has to be very careful not to stir up the surface layers. For the same reason core tubes with heavy weights attached have been found to be rather inefficient, as the bow wave of the tube blows away large numbers of meiofauna before the core strikes the sediment. Care should be taken, therefore, to check all possible sources of sampling error.

The sample sizes given above are practical compromises between a range of competing demands for statistical precision, ease of manipulating often heavy gear on a rolling and pitching ship, and the time taken to sort large amounts of sediment. However, it should be remembered that no single sample size is appropriate for all quantitative ecological studies. Just because a grab is available that samples 0.1 m^2 does not mean that that is the correct sample size for the community or species under study. In general a large number of small samples are preferable to a small number of large samples, since for the same counting effort a greater spread of habitats can be covered and the number of degrees of freedom for statistical tests is increased thereby reducing the error variance. However, if sample size becomes too small, then edge effects of the sampler become important. So sample size is a compromise between all of these factors. Elliot's book (1971) has a comprehensive account of how to determine sample size.

Similarly, the number of samples to be taken is often a compromise. The following example illustrates the sort of practical problems that occur. Ecologists are often interested in an estimate of how many species are in a given area and thereafter how many samples are needed to sample the community adequately. Fig. 1.5 shows a typical species–area curve obtained by pooling successive samples and determining the number of species in the pooled sample. The figure shows that within only two samples the number of species will vary from 15 to 25 and that the mean of five pairs of samples only gives an estimated 60% of the total number of species. Seven grab samples, however, yield between 29 and 34 species and represent 90% of the total number of species in the area. The actual

Fig. 1.5. Species–area curve for the macrofauna in the Helgöland Bight, Germany. Each point is the mean of five samples; bars show the range. (After Gerlach, 1972.)

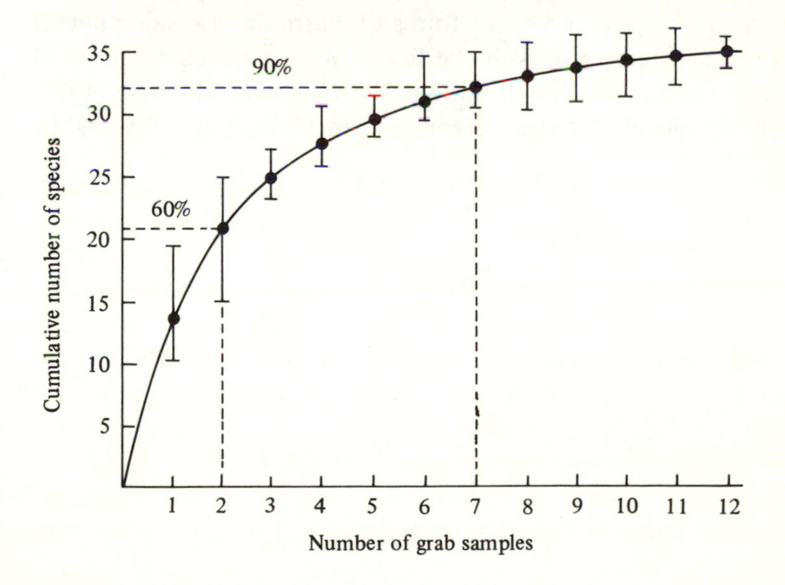

number of samples taken will be a compromise between the best possible representation of species and the time taken to obtain and process the samples.

Sampling strategy will often be dictated by the dispersion patterns of the fauna. With replicated samples a simple method of assessing the dispersion pattern of a species is to plot the variance : mean ratio. Values of less than 1 indicate a regular distribution and can be found in some territorial species. Values of around 1 indicate a random pattern, which seems to be common in deep-sea species at the scale of sampling used and also in the common intertidal bivalve, *Tellina tenuis*. Values of greater than 1 indicate contagious or aggregated distributions and are by far the most frequent patterns found in marine benthic species. The mention of scale with the deep-sea data is important, because dispersion patterns should ideally be investigated at many different sample sizes; they rarely are, however, since one is constrained by the type of sampling gear available. Quantitative deep sea samples with a large box-corer (0.25 m^2) show random patterns, but there may be aggregation patterns on a larger scale, since the food is hardly likely to be patchy in the deep sea. This point will be covered later (Chapter 6).

Since this book is not intended to be a practical guide, the methods for extracting and sorting the fauna will not be covered in detail. In general, various forms of elutriation (washing out) of the animals are used, with the fauna being retained on sieves of various sizes. Details can be found in Holme & McIntyre (1971) and, for meiofauna and microfauna, in Hulings & Gray (1971).

2

The sediment and related environmental factors

How can we characterise the sediment as a habitat for animals? The most obvious varying character is the grain size. On a typical sandy beach the coarsest particles lie at the top of the beach and grade down to the finest sediments at the water line. The top of the beach is dry and there is much windblown sand, since coarse sands drain rapidly, whereas at the lower end of the beach the sediments are wet, with frequent standing pools. Coarse sediment is found at the top of the shore because as the waves break on the beach the heaviest particles sediment out first. Finer particles remain in suspension

Fig. 2.1. The relationship between particle size and current speed for various factors affecting grain size distribution. Roughness velocity is the water velocity at which laminar flow changes to turbulent flow across the bottom. Threshold velocity is the current speed just sufficient to move a particle. Settling velocity is the settling rate of particles according to Stokes's Law. (After Inman, 1949.)

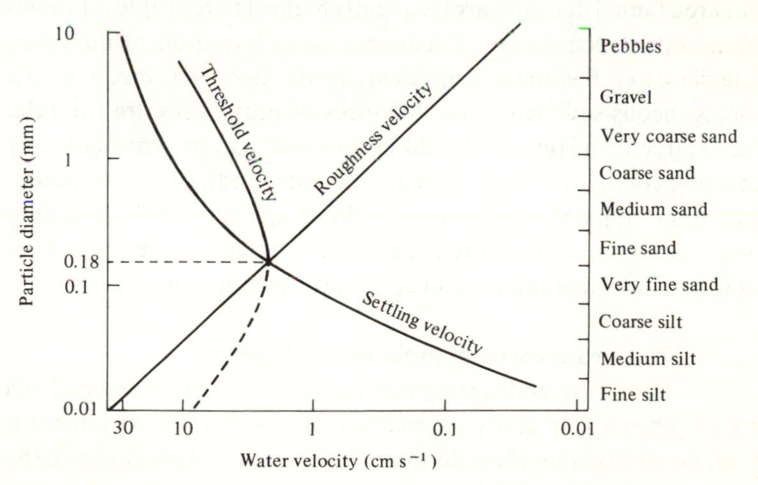

longer and are carried seaward on the wave backwash. Subtidally waves are important in distributing sediments down to depths of up to 100 m, but the dominant subtidal influences on sediment transport are currents. The type of deposit found depends on many factors, including current speed, the roughness of the sediment and the length of time when conditions are still and particles can sediment out. Fig. 2.1 shows the relationships between the important factors influencing the mobility of sediment particles. The curious fact is that particles of 0.18 mm diameter are the easiest to move. Coarser particles are difficult to pick up and transport since they are dense, whereas particles finer than 0.18 mm pack into a smooth bottom-surface and are difficult to resuspend. Thus if a sediment is composed largely of particles around 0.18 mm diameter it can be expected to be the most stable of all, since where such sediments occur wave and current action must be minimal. Clearly then, grain size, in the form of the median particle diameter, is a key character to measure.

In general, coarse intertidal sediments drain fast and retain little water or organic matter, and are therefore inhospitable habitats. On the other hand, fine sediments such as muds, which have grains tightly packed together, preclude the presence of an interstitial fauna and have poor water circulation and often low oxygen tension. Medium and fine sands usually have an abundant meiofauna and macrofauna, but because muds have more organic matter per unit area faunal densities are frequently highest here. Table 2.1 shows the accepted definitions of sediment types based on median particle diameter. Sediment samples are never, however, made up of a homogeneous sediment type. Mixtures of grain sizes are the rule. The degree of mixing of the different types can be represented by a sorting coefficient. Well-sorted sediments tending towards homogeneity are typical of high wave and current activity (high-energy areas), whereas poorly sorted sediments are heterogeneous and are typical of low wave and current activity (low-energy areas).

2.1 The measurement of grain size and sorting

Grain size and sorting can vary over distances as small as a few centimetres. In studying meiofaunal distribution patterns it is often necessary, therefore, to record the fauna and grain size on the same sample. With macrofaunal studies a single sample of 50–100 g from a grab is widely used for assessing grain size.

The exact details of grain-size measurements are beyond the scope of this book and only a general outline will be given; the interested reader is referred to Buchanan (1971) or to a fine geological book of methods by Robert L. Folk (1968). The silt–clay fraction is usually separated from the sand by use of 0.062 mm screen, since different methods are used for the analysis of sands (coarser than 0.062 mm) and of silts and clays (finer than 0.062 mm). Sands are dried and sieved on a series of screens that usually follow a decreasing geometric scale. The left-hand column of Table 2.1 shows the mesh size (in millimetres) of the usual range of screens used. By convention these are expressed on the phi (ϕ) scale, where $\phi = -\log_2$ of the particle size in millimetres. But to give sufficient points on plots of grain sizes a half-phi scale is recommended. Sediment finer than 0.062 mm is analysed by using a method based on sedimentation rates in cylinders containing tap water maintained at a constant temperature. Using Stokes's law, the settlement rate is calculated on a decreasing scale of particle size. The whole operation down to 10 phi (= 1 μm) takes over 16 hours to complete. The data obtained from such analysis are in the form of weights per screen. In fact, the

Table 2.1. *Grain-size characteristics*

Grain size (mm)	Phi (ϕ) scale	Type of sediment
256	−8	Cobble
64	−6	
16	−4	Pebble
4	−2	
2	−1	Granule
1	0	Very coarse sand
0.5	1.0	Coarse sand
0.25	2.0	Medium sand
0.125	3.0	Fine sand
0.0625	4.0	Very fine sand
0.031	5.0	Coarse silt
0.0039	8.0	Silt
0.002	9.0	
0.00006	14	Clay

distribution of particle sizes is of course continuous, but the data that are obtained are discrete.

Fig. 2.2 shows typical data from a series of screens for a sand sample, where the data are first expressed (Fig. 2.2*a*) as percentage dry weight of each fraction. Plotting this as a cumulative percentage gives an S-shaped curve if the data approximate a normal distribution (Fig. 2.2*b*). From this the median particle diameter can be obtained as the 50% point. Whilst it is easy to estimate the 50% point

Fig. 2.2. Illustration of grain-size plots for sediment from a sand beach. (*a*) Percentage dry weight against grain size; (*b*) cumulative percentage plot of the same data; (*c*) cumulative plot on probability paper.

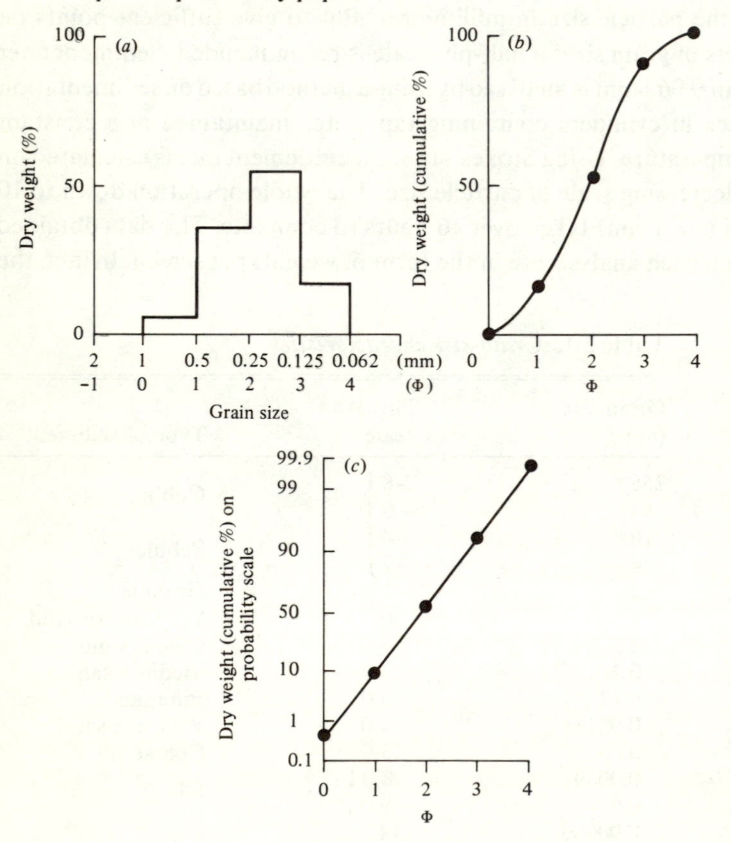

from Fig. 2.2(*b*), to obtain a sorting coefficient we need to know the phi values for 84% and 16%, and since the line is S-shaped it is difficult to read off these values accurately. Thus the curve is transformed into a straight line by plotting on a probability scale (Fig. 2.2*c*). One commonly used sorting coefficient is the Graphic Standard Deviation (σ_g). This is simply $\phi 84 - \phi 16/2$, so can be obtained from merely two values on the cumulative curve. A better index (which I prefer) is the Inclusive Graphic Standard Deviation (σ_I), given by the formula

$$\frac{\phi 84 - \phi 16}{4} + \frac{\phi 95 - \phi 5}{6.6}.$$

This formula covers over 90% of the distribution and is therefore a better overall measure of sorting. The sorting classes produced by this index are as follows:

Under 0.35 ϕ	very well sorted
0.35–0.50 ϕ	well sorted
0.50–0.71 ϕ	moderately well sorted
0.71–1.00 ϕ	moderately sorted
1.00–2.00 ϕ	poorly sorted
2.00–4.00 ϕ	very poorly sorted
Over 4.00 ϕ	extremely poorly sorted

Whilst grain size and sorting are probably the two most important characters that can be measured on sediment samples, other biologically important properties include porosity and permeability, which are particularly relevant for meiofaunal studies. Porosity measures the amount of pore space available and permeability the rate of percolation of water through the sediment. (Methods can be found in Holme & McIntyre, 1971.)

Wave action and current velocity are the two most important factors determining the grain-size distribution and sorting coefficients of nearshore sediments. In nearshore environments the organic content of the sediment typically increases with the fineness of the deposit, since particles of sedimenting organic matter behaves as do sediment particles (although, of course, their specific gravity is lower).

2.2 The measurement of organic content

The bulk of the organic material in sediments (excluding the fauna) is derived by sedimentation from the overlying water column. In nearshore areas over the continental shelf, where plankton production is highest, the organic content of the sediment is similarly at its maximum. (This topic will be discussed more fully in Chapter 11, where energy flow in sediments is dealt with.) The organic content is usually measured as organic carbon or nitrogen, and the C/N ratio is widely used. Today it is possible routinely to process sediment samples through a C-H-N analyser. In some sediments measurement of the organic carbon content can give a false impression of their potential productivity since the carbon is present as coal particles which cannot be utilised directly by bacteria. To overcome this problem one technique is to estimate the organic matter occurring as protein by extracting a number of amino acids (a method is described by Buchanan & Longbottom, 1970).

The organic material arriving at the sediment surface is broken

Fig. 2.3. The sulphur cycle in sediments. (After Ferguson-Wood, 1965.)

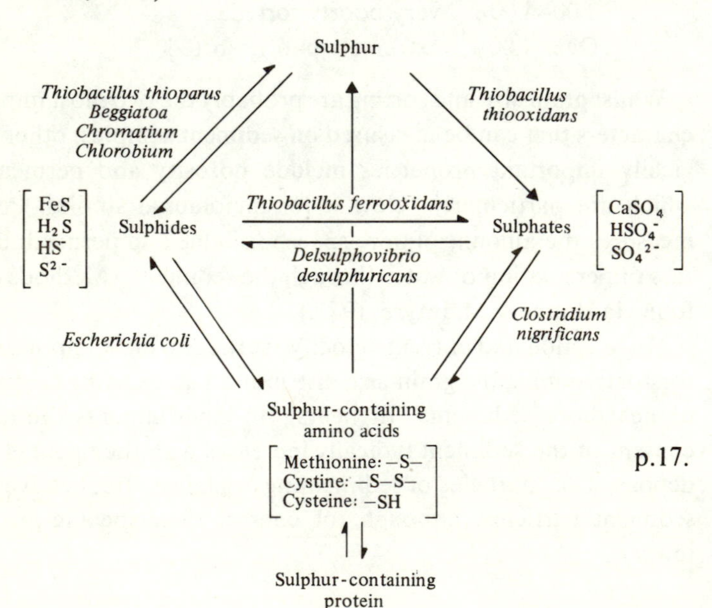

p.17.

down by bacterial action. In the marine environment, just as in the terrestrial, the carbon, nitrogen and phosphorus cycles operate. Primary productivity is limited by the availability of nitrogen rather than phosphorus and so the nitrogen cycle is very important (Ferguson-Wood (1965) has reviewed marine microbial ecology). Of particular importance in marine sediments, however, is the sulphur cycle. In poorly drained sandy beaches and in nearly all muddy sediments a black layer of reduced sediment with a smell of rotten eggs (hydrogen sulphide: H_2S) is common. Fig. 2.3 illustrates the sulphur cycle in the sediment. The protein from dead plants and animals is broken down into amino acids and then directly to sulphides or more usually to sulphates. The sulphate–sulphide reduction is the most important in the chemistry of sediments since the pH and redox potential (Eh) of the sediment are determined by the bacteria responsible for sulphate reduction and from this reaction hydrogen sulphide is produced. On intertidal mudflats the filamentous bacterium *Beggiatoa* is often common, with the purple bacterium *Chromatium*, or the green sulphur bacterium *Chlorobium*. These bacteria oxidise sulphides to elemental sulphur, which is deposited inside the cells of *Beggiatoa* and *Chromatium* but outside the cells of *Chlorobium*.

If the sediment is examined closely it will be found that there is an oxygenated brown layer which changes to a grey layer immediately above the black sulphide layer. This grey layer marks the transition between oxygenated and reduced conditions. The degree of oxygenation or reduction is measured by the redox potential (Eh) of the sediment. This is done by replacing the electrode of a pH meter by a naked platinum electrode. In oxygenated sediments Eh values can be up to + 400 mV; they go down to −200mV in strongly reduced sediments. Fig. 2.4 shows typical redox potential profiles for a medium sand and a fine sand. The layer of rapid change of redox potential with depth corresponds to the grey layer. This has been called the redox potential discontinuity layer (RPD layer), and is an important environmental factor to measure in sediments. In fine sediments (Fig. 2.4*b*) the RPD layer lies closer to the surface than it does when the sediment is coarser. In subtidal sediments measurement of redox potential can be made directly, by diving, or by measurement in undisturbed cores removed from the sediment.

Sulphide ions are toxic to nearly all aerobic species and so the RPD layer indicates an important lower limit of depth distribution for many species (unless the species possess tubes whereby oxygenated water can be drawn down from above).

But although the RPD layer indicates an effective ecological barrier for most species, not all organisms are absent in the black layers of sand beaches. Recently a whole group of specialised meiofauna has been found that occurs exclusively in this environment. This group of fauna has been called the 'thiobios' (Fenchel & Riedl, 1970). Among other remarkable adaptation the group appears not to have mitochondria, and it may be an extremely primitive fauna.

2.3 Other factors

The other commonly measured environmental factors – temperature, salinity and oxygen – present few problems. There are many commercially available thermistor probes ideal for measuring the temperature in sediment. Similarly, salinity can nowadays be measured on a single drop of water with an acceptable degree of accuracy by using refractometers. More accurate estimates of salinity can be made by taking water samples for conductivity measurements, but the large volumes of water needed are difficult to obtain when the interest is in small-scale vertical and horizontal salinity gradients. Oxygen electrodes are not usually suitable for direct

Fig. 2.4. Typical redox potential profiles for two intertidal sediments: (*a*) medium sand, (*b*) fine sand.

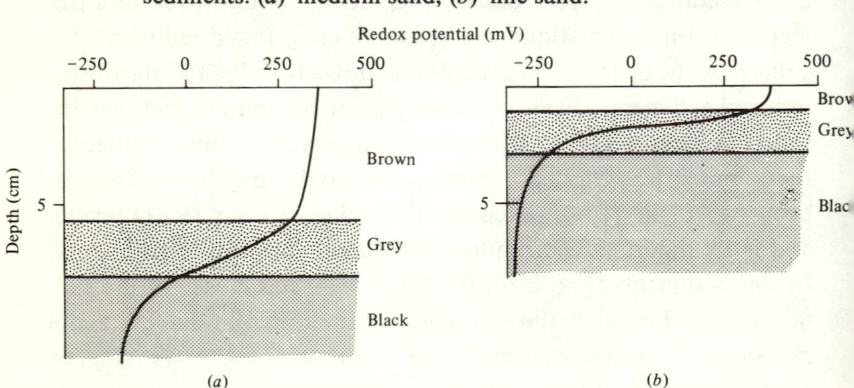

measurements in sediment, both because of the thin, easily damaged membranes that cover the probe and also because of the risk of poisoning by hydrogen sulphide. The most widely used method for measuring oxygen concentration is to take a water sample and use a micro-Winkler technique.

In intertidal and subtidal sediments where there is abundant light, there is considerable *in situ* primary production. Measurement of biomass from estimates obtained by extracting chlorophyll pigments is probably the commonest method used (see Hickman & Round (1970) for a detailed account). Production has been measured using a modified version of ^{14}C method of phytoplankton studies (Hickman & Round (1970) give methods). In production and energy flow studies (see Chapter 11) it has been found that primary production of sediment can be a highly important source of food for the benthos. There are, however, relatively few studies that have dealt with this important topic.

2.4 Seasonal variation in environmental factors

Having given an outline of how the various environmental factors can be measured, some indication of the sorts of seasonal changes that can be expected must be given. In the winter, when there are heavy seas and therefore more erosion, the beach is coarser; in summer there is a build-up of fine particles that leads to a reduction in median grain size and sorting. As a consequence, the organic content rises and the RPD layer moves closer to the surface. Temperatures may well range over the year from −5 °C to +30 °C, and a cloud-burst can cause the salinity at the surface to change from 35‰ to less than 10‰ in a few minutes. The depth to which salinity changes reach is surprisingly small, however, and indications are that below 10 cm salinity is virtually constant. Changes in intertidal sediments are much greater than those occurring in subtidal sediments. The increased fineness of the deposit in summer and the rise in the RPD layer are general phenomena, but variations in temperature and salinity are much less marked subtidally.

3

The distribution of individuals among species

Let us assume that a sample has been taken from a community, the environmental variables measured and the species identified and individuals counted. What approaches can be taken to analyse the data?

Any field biologist is constantly amazed by the complexities of nature and especially the marvellous adaptations that species show to their habitat and to other species. One of the most fruitful aspects of ecological research, however, is the search for common patterns in this bewildering variability. From any sample of a biological community, whether marine, terrestrial or freshwater, there is an immediately observable pattern that a few species are very common, represented by many individuals, whilst most species are rare, represented by one or a few individuals. The obvious question is whether any 'rules' can be applied to these patterns.

There are two common ways of expressing such patterns: as a rank of abundance from the most to the least abundant species, or as a frequency distribution of the individuals per species plotted against the number of species.

3.1 Ranked abundance models

The simplest of all patterns is that obtained by plotting relative abundance as a percentage on a logarithmic scale against species rank. Fig. 3.1(*a*) shows a typical plot of, in this case, the benthos of a highly polluted part of Oslofjord. This distribution is called the geometric series, and here the dominant species constitutes a very large proportion of the whole community. The data which best fit such a series are those from early successional stages or from harsh environments. In both of these cases a high degree of dominance results and the distribution of individuals per species is very uneven. From this pattern a model called the niche pre-emption model has been suggested.

3.1.1 *The niche pre-emption model*

This model applies to simple communities composed of only a few species from taxonomically similar groups where the habitat is fairly uniform. It is envisaged that the most successful species occupies (pre-empts) the largest share (K) of the limiting resource (usually space in benthic communities). The next most successful species pre-empts the next largest share, and so on. The result is that the ranked abundance list forms a geometric series.

Fig. 3.1. Models of distribution of individuals among species for marine benthic data (●—●, observed; △—△, expected). (*a*) Geometric series from benthos of a heavily polluted part of Oslofjord, Norway (data of F. B. Mirza). (*b*) Broken-stick model of ophiuroids from Eniwetok (data from King, 1964). N, total number of individuals; s, total number of species. (*c*) Log-series distribution from meiofauna of kelp holdfasts (data from Moore, 1973).

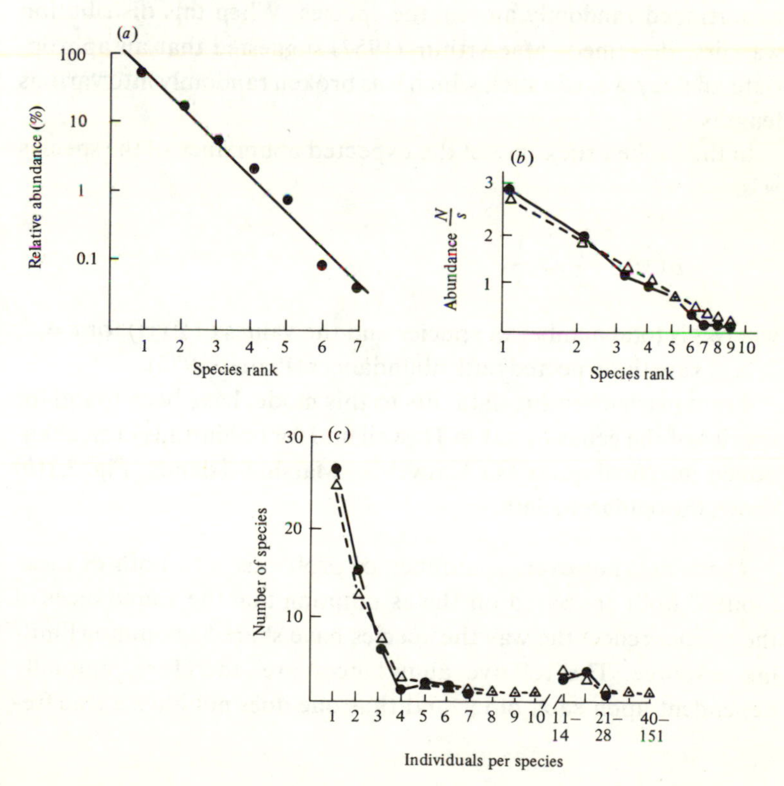

In this model it is assumed that the abundance of each species is proportional to the resource fraction that it pre-empts. The ranked abundance list then becomes

$$K, K(1-K), K(1-K)^2, \ldots, K(1-K)^{s-2}, (1-K)^{s-1},$$

where s is the total number of species (Pielou, 1975).

Another model, based not simply on plotting the data but on having a calculated expected distribution is the broken-stick model.

3.1.2 *The broken-stick model*

This model from ranked abundance lists was originally also based on a simple community of a few species from taxonomically similar groups in a uniform habitat, but recently Pielou (1975) has shown that the restriction of this model to simple communities from a homogeneous habitat is not necessary. Species are envisaged as competing for a major limiting resource so that the resource is apportioned randomly among the species. When this distribution was first described, MacArthur (1957) suggested that an appropriate analogy was of a stick which was broken randomly into various lengths.

In the broken-stick model the expected abundance of the species yi is:

$$E(yi) = \frac{1}{s} \sum_{x-1}^{s} \frac{1}{x},$$

where s is total number of species and the values of $E(yi)$ for $i = 1, 2, \ldots, s$ are the expected rank abundances (Pielou, 1975).

From marine benthic data, fits to this model have been found for species of the genus *Conus* in Hawaii and for ophiuroids in an abandoned intertidal quarry in Eniwetok, Marshall Islands, Fig. 3.1(*b*) shows the ophiuroid data.

There are, however, a number of problems with both of these models. Both are based on the assumption that the abundances of the species reflect the way the species have shared a common limiting resource. The relative abundances are, therefore, mutually dependent upon each other and thus one does not have a true frequency distribution since one single value determines the whole pattern of the data.

Pielou (1975) has pointed out that the basis of these models is two biological assumptions. The first of these is that, following competition on a local scale the resource has been divided among the species and results in the given model. The second assumption is that over evolutionary time the species have become adapted to different tolerance ranges and the relative sizes of these tolerance ranges give the particular model. In the former case the division of the resource gives the realised niche of the species, whereas in the latter case the outcome is the fundamental niche of the species. (See Chapter 5 for a fuller explanation of the realised and fundamental niche). The niche pre-emption model will presumably always predict fundamental niches, whereas with the broken-stick model it is not clear which assumption is used.

The niche pre-emption model is found so seldom in nature that as a general model it seems ecologically unrealistic. The problems with the broken-stick model are equally great. Firstly, it can be derived from a simple mathematical distribution (see Pielou, 1975), whereas MacArthur's original idea in producing the model was that it had a biological basis. Furthermore, in the broken-stick model all possible ranked abundance lists are equally probable, so that the relative abundances in a given community can neither give support to the model nor disprove it. This may seem a little hard to understand, but Pielou has given a simple analogy. In a shuffled pack of cards a randomly drawn card could have any value from 1 to 13. If one repeats the random withdrawals an infinitely large number of times, the expected score will be 7, since this is the mean value. The so-called 'expected' value of the broken-stick model is simply the mean of the infinitely large number of repetitions. Therefore, any species abundance distribution is equally likely as any other. In fact as long ago as 1966 MacArthur abandoned the broken-stick model as not being ecologically useful, but it is still frequently referred to.

3.2 Frequency distributions

Another way of expressing the distribution pattern of individuals among species is by plotting individuals per species against number of species. When this was first done it seemed that the pattern was that the commonest class was those species represented by one individual, the next commonest class those represented by two individuals, and so on. Such patterns have been found in a number of sets of benthic data (Fig. 3.1*c*).

The great English statistician R. A. Fisher realised that such a pattern conformed to a logarithmic series.

Fig. 3.2. Derivation of the log-normal distribution of individuals among species for macrobenthos of a Scottish loch (data from Pearson, 1975). (*a*) Raw data showing skewed distribution; (*b*) approximately normal distribution on a × 2 geometric scale (see Table 3.1).

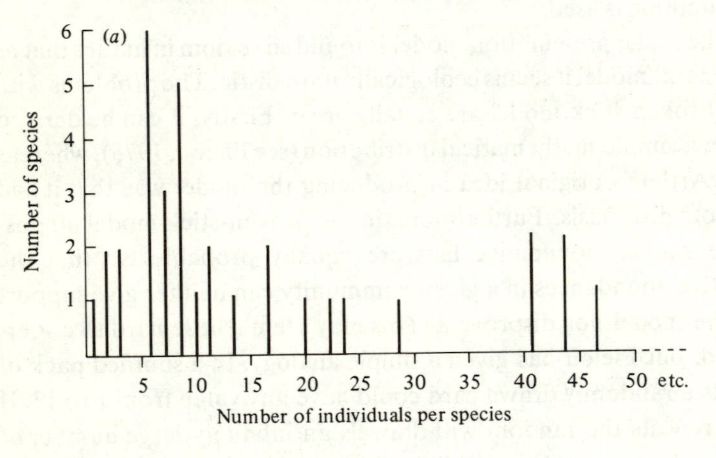

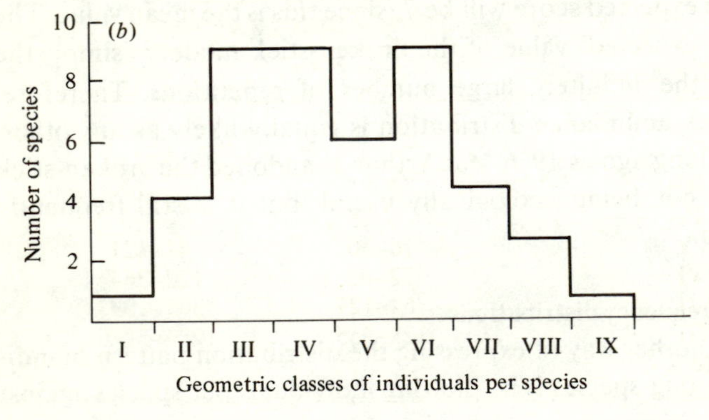

3.2.1 *The logarithmic series*

The terms of the log series can be written:

$$\alpha x, \alpha\frac{x^2}{2}, \alpha\frac{x^3}{3}, \alpha\frac{x^4}{4}, \ldots,$$

where αx is the number of species represented by one individual, $\alpha\frac{x^2}{2}$ the number of species represented by two individuals, etc. Knowing the total number of species and the total number of individuals in the community α can be calculated (or found from a table in Williams, 1964). The goodness-of-fit of the empirical data to the model can then be tested using a chi-square test. In addition to the kelp holdfast data shown in Fig. 3.1, the log series has been found to fit data on the benthos off the coast of Northumberland and on meiofauna from Yorkshire (Gray, 1978).

In many samples of benthic communities, however, the most abundant class is not that represented by one individual per species, but often lies between classes with three and those with six individuals per species (Fig. 3.2). In this case the appropriate frequency distribution is the log-normal distribution.

Table 3.1. *Relationship between the arithmetic and geometric scales used in plotting the log-normal distribution of individuals among species*

	Arithmetic scales: (individuals per species)	
Geometric scale	× 2	× 3
I	1	1
II	2–3	2–4
III	4–7	5–13
IV	8–15	14–40
V	16–30	41–121
VI	32–63	122–364
VII	64–127	365–1093
VIII	128–255	1094–3280
etc.		

3.2.2 *The log-normal distribution*

If one plots the number of individuals per species against the number of species, the curve obtained is frequently strongly skewed (Fig. 3.2*a*). This curve can be brought to the familiar normal curve by plotting the number of individuals per species on a geometric scale. The geometric scale recommended by Preston (1948), who was the first to apply the log-normal distribution to ecological data, was a × 2 scale (Table 3.1). A × 3 scale is often used also.

Fig. 3.2(*b*) shows the same data as in (*a*) but plotted on a geometric scale of individuals per species. The curve is now a normal one, but on a geometric scale, hence the log-normal distribution. Plots of the log-normal type typically do not show the complete curve, but are truncated at the left-hand side because not all the species were found. Had a larger sample been taken then more of the truncated part of the curve would be exposed. This provides a method for arriving at an estimate of the species that were not caught. In order to make this a relatively easy operation one can utilise the fact that plots of a normal curve on probability paper give a straight line if the species are plotted as a cumulative percentage. Thus a log-normal distribution will produce a straight line if cumulative species percentages are plotted against a geometric scale of individuals per species. Extrapolation of the line back to the *y*-axis gives a rough estimate of the percentage of species not sampled. Fig. 3.3 shows data from three different benthic communities that are an excellent fit to the log-normal distribution. Many similar examples could be plotted. In each of the examples shown the sampling was extremely good, since extrapolation back to the *y*-axis shows that only just over 5% of the number of species were not sampled. Often this is not the case and the figure can be as high as 30% or more.

Whilst the method above is a quick and easy method for testing the goodness-of-fit to a log-normal distribution, there is a more accurate statistical method. For the exact fitting of both the log-series and log-normal distributions see the authoritative account by Bliss (1966).

Fig. 3.3 shows that the log-normal distribution is an accurate description of benthic samples. It has also been shown to be a good fit to communities as diverse as moths in England, snakes in Panama, birds in New York State and in Finland, and phytoplankton in the

North Atlantic – in fact, almost any community that has been studied. The only provisos are that the communities must be heterogeneous and the samples large. Why should this distribution be such a good fit?

Firstly, populations tend to increase geometrically rather than arithmetically. Thus when a large number of species are considered together one would expect the number of individuals per species to be distributed in a geometric manner. Secondly, the distribution

Fig. 3.3. Log-normal distributions of individuals per species. Geometric classes on × 2 scale (see Table 3.1). (*a*) Meiofauna of kelp holdfasts (data from Moore, 1973); (*b*) macrofauna from Irish Sea (data from Eagle & Hardiman, 1977); (*c*) macrofauna of a Scottish sea loch (data from Pearson, 1975).

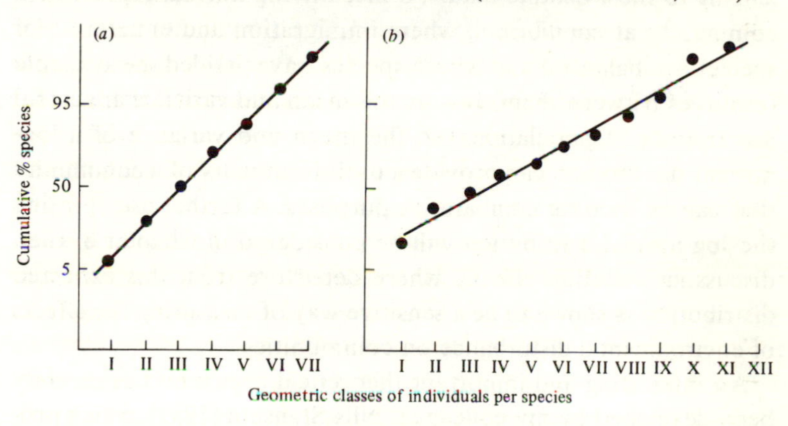

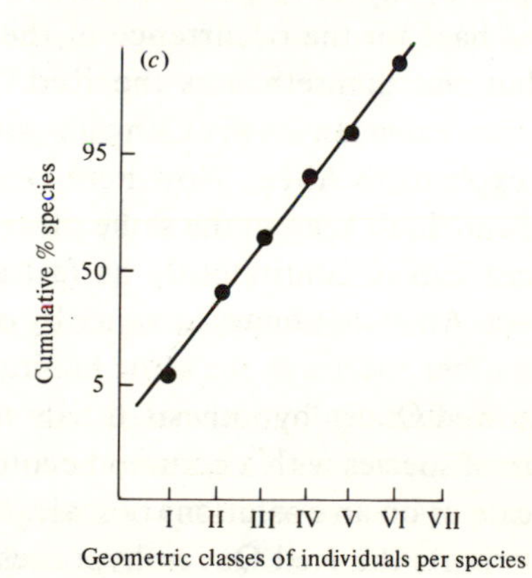

of individuals among species in a community can be considered the result of the interaction of many environmental factors each acting randomly. The combined action of these many factors acts multiplicatively rather than in an additive manner, and the Central Limit Theorem applied to the product will give a log-normal distribution (May, 1975). Furthermore, one can derive the log-normal distribution by postulating that in the broken-stick model breakages occur sequentially rather than at random, so that the probability that a part will be broken is independent of its length (Pielou, 1975).

The log-normal distribution is, then, to be expected for large and heterogeneous assemblages of species and is a good description of the distribution of individuals among species without necessarily implying any biological significance to the fit. Indeed it is an excellent fit to most benthic data. To me, the log-normal represents a community at equilibrium, where immigration and emigration of species are balanced and where species have divided the available resources between them. Just as the mean and variance are useful descriptions of populations, so the mean and variance of a log-normal distribution can provide a useful summary of a community that can be used for comparative purposes. A further use of fitting the log-normal distribution will be considered in Chapter 8 when discussing pollution effects, where departure from this expected distribution is shown to be a sensitive way of measuring the effects of environmental disturbance on communities.

An interesting and important theoretical argument has recently been developed by my colleague, Nils Stenseth (1979), which provides a sound base for the occurrence of the log-normal and log-series distributions. Stenseth uses the Red Queen hypothesis put forward by Van Valen. In Lewis Carroll's *Alice in Wonderland* the Red Queen explains to Alice, 'Now here, you see, it takes all the running you can do to keep in the same place'. The analogy is that a species must evolve continuously as fast as it can to avoid becoming extinct. An evolutionary advance by one species has a negative effect on other species in the same habitat. Van Valen suggests, based on the Red Queen hypothesis, a 'law of constant extinction' for any group of species with a common ecology. Extinction can be on a local scale or on an evolutionary scale. Stenseth's mathematical model supports the Red Queen hypothesis and predicts a constant diversity over ecological and evolutionary time: also that species abundance curves will be log-normal in undisturbed communities but that a log-series distribution will fit a disturbed community. He provides empirical support for the predictions. Another consequence of the Red Queen hypothesis is that the area relationships of species have a set mathematical pattern and, therefore, that there is a maximum number of species that a given area can support. Stenseth sums up his paper thus: 'On the basis of the Red Queen hypothesis regulating mechanisms are likely to be related to how good any particular species is relative to all others in utilising the limited amount of resources; what is important in order not to become extinct, is to evolve as fast as you can. Those that lag behind become extinct; this sets the limit to how many species there can be in any community.'

4

Classifying assemblages of species

In the previous chapter the patterns of the distribution of individuals among species were considered, but the identity of the species was not used. The next type of pattern that can be sought is that between groups of species that are found in similar places. Here one is interested in finding out how groups of species occur together and in trying to divide up the long lists of species and stations into more homogeneous groups. Are there in fact communities of similar species occurring together over large areas of the sea-bed? The traditional way of classifying assemblages of species on rocky shores has been to use the dominant species. With benthic communities the classical work of Johannes Petersen (1914, 1915, 1918, 1924), using a similar idea, laid the foundation for a scheme that is still widely used today.

4.1 Traditional methods: the Petersen–Thorson system

Working predominantly in the Kattegat around the Danish islands of Fyn and Zeeland, but also extending his work into the Skaggerak and the North Sea, Petersen sampled the shallow-water benthos quantitatively, recording species present and their numbers and weights (Fig. 4.1 shows his sampling stations). In all, he worked at 193 sites and listed 294 faunal records. Not all were recorded to the level of species (of which there were 260), but were recorded as juveniles and adults separately. Petersen designated a series of communities by their so-called characterising species. A characterising species was one which was not seasonal and which, owing to numerical or biomass dominance, could be regarded as typical of a given assemblage. Constancy and dominance were, therefore, the two most important characters. In all Petersen recognised seven major communities, characterised by the following: (1) the bivalve *Macoma balthica*; (2) the burrowing echinoderm

Fig. 4.1. Petersen's sampling stations around Denmark. In addition he sampled on the Swedish coast and in the Oslofjord (not shown).

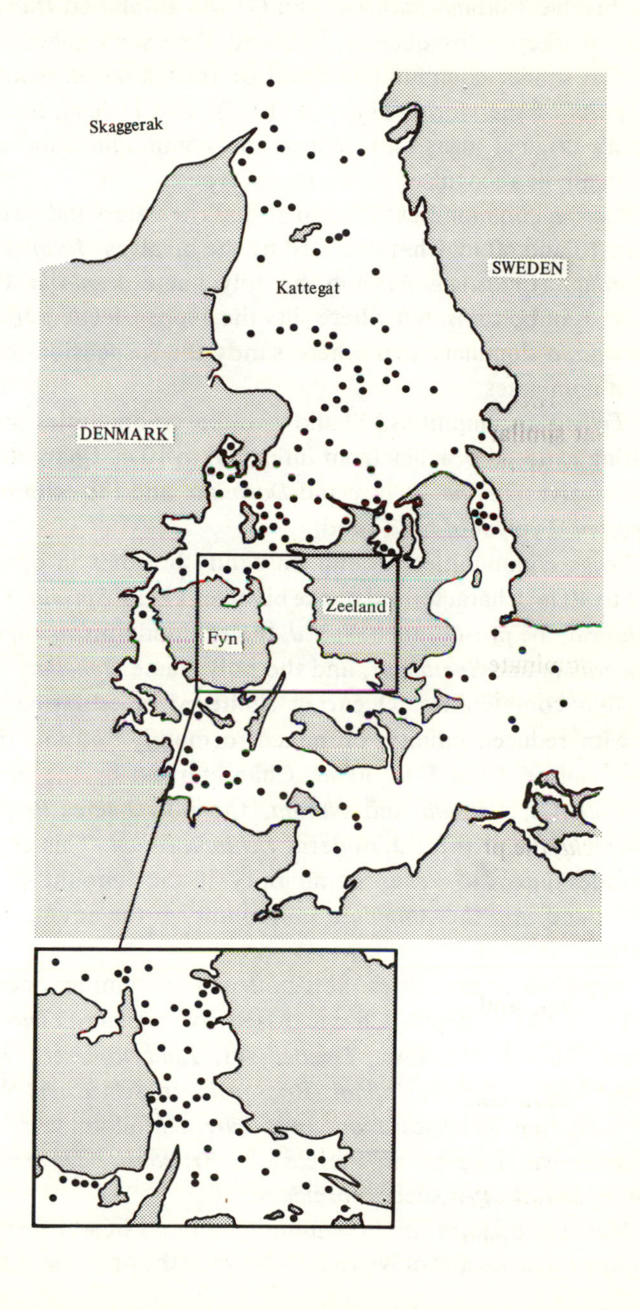

Brissopsis; (3) the burrowing echinoderm *Echinocardium*; (4) the bivalves *Asarte, Abra* and *Macoma calcarea*; (5) the bivalve *Venus*; (6) the bivalve *Macoma calcarea*; and (7) the amphipod *Haploops*.

Many workers subsequently followed Petersen's scheme and found that similar communities could be recognised in many different parts of the world. Thorson (1957), also Danish, extended Petersen's original ideas and defined the communities more precisely. In all, he recognised seven major types.

(1) *Macoma* communities. Typical of shallow waters and estuaries between 10 and 60 m. Characterised by the bivalves *Macoma, Mya* and *Cardium* (*Cerastoderma*) and the polychaete *Arenicola*. Occur on all types of bottoms, but where silty the deposit-feeders *Macoma* and *Arenicola* dominate and where sandy the suspension-feeding *Cardium* dominates.

(2) *Tellina* communities. Shallow-water communities mainly inhabiting exposed beaches from intertidal to 10 m. Characterised by the bivalves *Tellina, Donax* and *Dosinina*, and the echinoderm *Astropecten*. Typical of hard sand.

(3) *Venus* communities. Found on sandy bottoms in open sea from 7 to 40 m. Characterised by the bivalves *Venus, Spisula, Tellina* and *Thracia*, the prosobranch *Natica*, the echinoderms *Astropecten, Echinocardium* and *Spatangus*, and the polychaete *Ophelia*.

(4) *Abra* communities. Occur in sheltered or estuarine areas, often with reduced salinity, on mixed to muddy bottoms rich in organic matter, from 5 to 30 m. Characterised by the bivalves *Abra, Cultellus, Corbula* and *Nucula*, the polychaetes *Pectinaria* and *Nephthys*, and the echinoderm *Echinocardium*. This community grades into the *Venus* community if the amount of sand increases and into the *Amphiura* community if the amount of silt increases.

(5) *Amphiura* communities. Soft-bottom communities occurring from 15 to 100 m depth. Characterised by *Amphiura, Turritella, Thyasira, Nucula, Nephthys, Terebellides, Limbriconereis, Dentalium* and one of *Echinocardium, Brissopsis* or *Schizaster*. With a sandy substrate *Echinocardium* and *Turritella* dominate, while more mud will lead to an increase in *Brissopsis, Thyasira* and sedentary polychaetes such as *Maldane*.

(6) *Maldane–Ophiura sarsi* community. Found in soft fine muds in shallow estuaries and down to 100–300 m in the open sea. Charac-

terised by the polychaetes *Maldane* and *Terebellides*, the echinoderm *Ophiura sarsi*, the bivalves *Nucula, Abra* and *Thyasira*, the gastropod *Philine*, the polychaetes *Aricia*, *Melinna, Praxilella, Clymenella, Glycera* and *Pectinairia*, the amphipod *Ampelisca* and the echinoderms *Brissopsis* and *Echinocardium*.

(7) Amphipod communities. Estuarine or brackish-water communities usually on soft bottoms. Characterised by various amphipods each typical of its own community: e.g. *Pontoporeia* in the Baltic, *Haploops tubicola* in some areas of Denmark, *Ampelisca* in Japan and Massachusetts.

Thorson had studied communities in many parts of the world and was struck by the fact that while the same genera often occurred on the same types of bottom in different areas, the species were different on going from the artic-boreal regions to, say, the Pacific coast of North America. He called these communities parallel communities. For example, the *Macoma* communities could be split up into four parallel communities: the *M. calcarea, M. balthica, M. nasuta–M. secta* and the *M. incongrua* communities. The first two are found in the North Atlantic, occurring in east Greenland and most of Europe respectively, whereas the *M. nasuta–M. secta* community is typical of the north-east Pacific in the San Juan archipelago of Washington State, USA. The *M. incongrua* community is the typical north-west Pacific community of Japan. Similar parallel communities were recognised by Thorson for the *Tellina* and *Venus* communities.

In the late 1950s more research was concentrated on the benthic fauna of warmer waters and many reports were made that the Petersen–Thorson communities could not be identified. This was explained away by the fact that there were many more species in warmer waters and so dominance patterns similar to those in colder waters were not to be expected. Then in the 1960s an increasing number of reports appeared where communities could not be adequately defined even in the colder waters. Questions were raised as to the whole validity of the community concept.

4.2 Rigid communities or continua?

Remarkably enough, the same debate had been raging in terrestrial plant ecology, about two decades earlier! The botanists had been divided into two camps. Firstly, there were those who believed that it was possible to classify assemblages of species into communities and that there were sharp boundaries between adjacent communities corresponding to discontinuities in the habitat. This view corresponds roughly to the Petersen–Thorson ideas on classifying benthic communities. The other camp of botanists believed that rather than plant species occurring in discrete groups with sharp boundaries, species occurred along gradients of environmental factors, with each species having an optimum somewhere along the gradient. The species overlapped in distribution and there were no discrete boundaries; one 'community', therefore, graded into another 'community'. Fig. 4.2 illustrates diagrammatically the two ideas. Which of the two is most appropriate for benthic communities?

Fig. 4.2. (*a*) The community concept: species distributions have sharp boundaries, often corresponding to discontinuities in the environment. (*b*) Species form continua along environmental gradients and occur in log-normal abundance patterns.

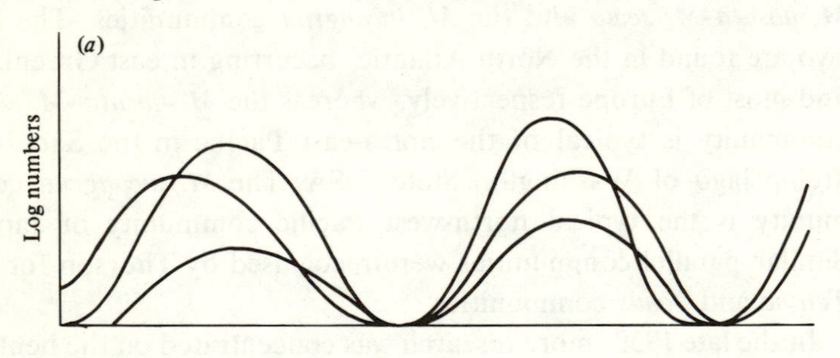

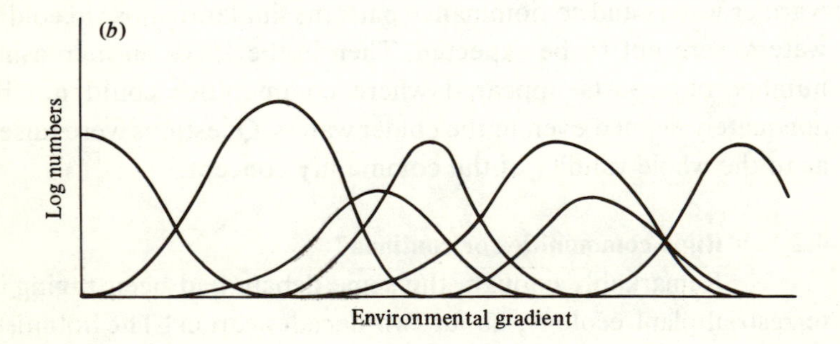

There seems to be little doubt that most recent data favour the idea that species are distributed in the form of continua. A good example of this is provided by the study of Hughes & Thomas (1971) from an intertidal beach in Canada. Fig. 4.3 shows their data, which indicate that species are distributed in approximately normal distributions along the gradient of distance (probably reflecting sediment grain size). The gradient could just as easily be salinity, in which case there would be another pattern of species. If the *y*-axis is plotted on a logarithmic scale of numbers, then the distributions are even more typically normal. Thus species seem to be distributed in log-normal curves of abundance along environmental gradients, i.e. do not form discrete communities. From these ideas stems a modern definition of communities (by Mills, 1969) which has met widespread acceptance: 'community means a group of organisms occurring in a particular environment, presumably interacting with each other and with the environment, and separable by means of ecological survey from other groups'.

Fig. 4.3. Distribution curves of macrofaunal species along an oyster bed in Canada: 1, *Mya arenaria* (× 2); 2, *Nereis virens* (× 10); 3, *Nassarius obsoletus*; 4, *Zostera marina* (× 10); 5, *Neopanope texana* (× 20). (Data simplified from Hughes & Thomas, 1971.)

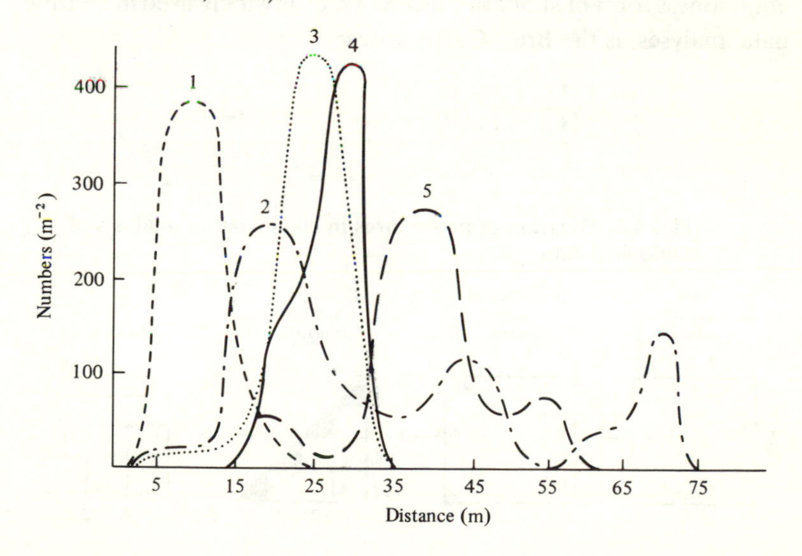

Both Petersen and Thorson selected the species that they intuitively thought were the most appropriate to indicate a certain community, and their choice was often highly subjective. A computer, on the other hand, will only sort data on specific criteria given in advance by the programmer. Computer-based sorting techniques are often called objective, but in fact much of the objectivity disappears on close examination of the methods used, since species are subjectively eliminated, for example, before the data are submitted to the computer. It is a surprising fact to people not used to computers that a common ecological data-set of 150–200 species from 200 samples is 'too large' for the computer. A further problem of computer sorting arises because most data-matrices contain many zero entries and, in accordance with the log-series and log-normal distributions, most species are rare, being represented by only one or two individuals. Too many zeros and ones skew the patterns obtained and render interpretation difficult, so a common practice is to delete all species that occur in less than, say, 5 of 70 sample sites.

One of the best general computer-sorting techniques is classification analysis. Fig. 4.4 illustrates diagrammatically the stages in the analysis. First the raw data-table of n samples and s species containing abundance data is transformed to the form log $(x + 1)$. From this transformed data a similarity matrix is derived. The commonest form of similarity index, which is widely used in benthic data analyses, is the Bray–Curtis index:

$$D = \sum_{i=1}^{s} \left| \frac{x_{1j} - x_{2j}}{(x_{1j} + x_{2j})} \right|,$$

Fig. 4.4. Diagram of procedures in classification analysis of ecological data.

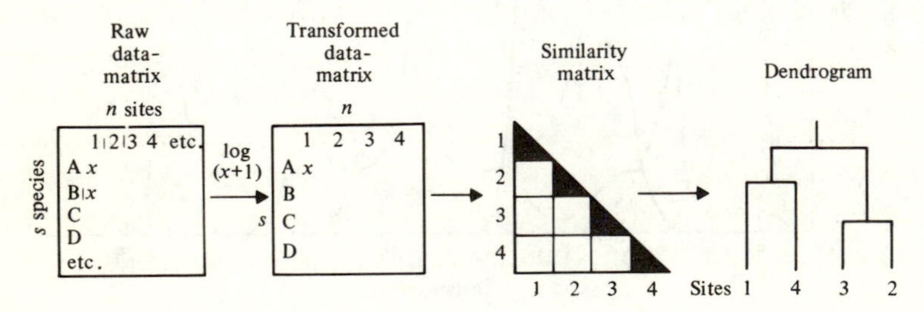

where x_{1j}, x_{2j} are the abundances of species j in sites 1 and 2, and s is the number of species. From this matrix a dendrogram is prepared showing the relationships between groups of samples. The group-average sorting strategy is widely used for this. Fig. 4.5 shows a dendrogram presenting the classification of 68 sampling stations from F. B. Mirza's study of the benthic macrofauna of the Oslofjord. A total of 146 species were found, but the matrix was prepared on only 69 species. In the figure the groupings to the left are odd samples usually poor in fauna, which are in many ways atypical. To the right of centre of the dendrogram group A includes all those stations that contain large numbers of *Capitella capitata, Polydora ciliata* and *Heteromastus filiformis*, species which are common in highly organically enriched sediments. Other groupings of stations (B, C and D) show fauna characteristic of only slight pollution (see Chapter 8 for a fuller discussion of these aspects). Thus the classification analysis has divided up the stations into groups containing similar species. Plotting the groups back onto a map of the fjord (Fig. 4.6) indicates patterns of similar stations. These show that the most polluted area (that containing group A species) occurs in the innermost part of the Oslofjord, the Bunefjord, where the bulk of the sewage from the city is discharged. As a follow-up to this work it is planned to study the dynamics of the communities within each of the major groups depicted in Fig. 4.6. Thus the computer analysis is not an end in itself, but suggests hypotheses which can be investigated, in this case the gradient of organic pollution.

There are a number of other multivariate computing techniques for use with large ecological surveys, one of which is ordination. Here the idea is that the variance in the data is decomposed into derived factors. Groupings of stations are then produced and ranked along the axes of two or three factors. Finally, one attempts to explain what the factor might be in terms of some environmental measurements, which were not used in the analysis but can be correlated with the factor. In practice the interpretation of the factors is hardly ever successful and environmental variables rarely seem to correlate directly with any factor, so that the method generally is not so effective as classification analysis. A thorough account of both classification methods and ordination techniques is given in Clifford & Stephenson (1975).

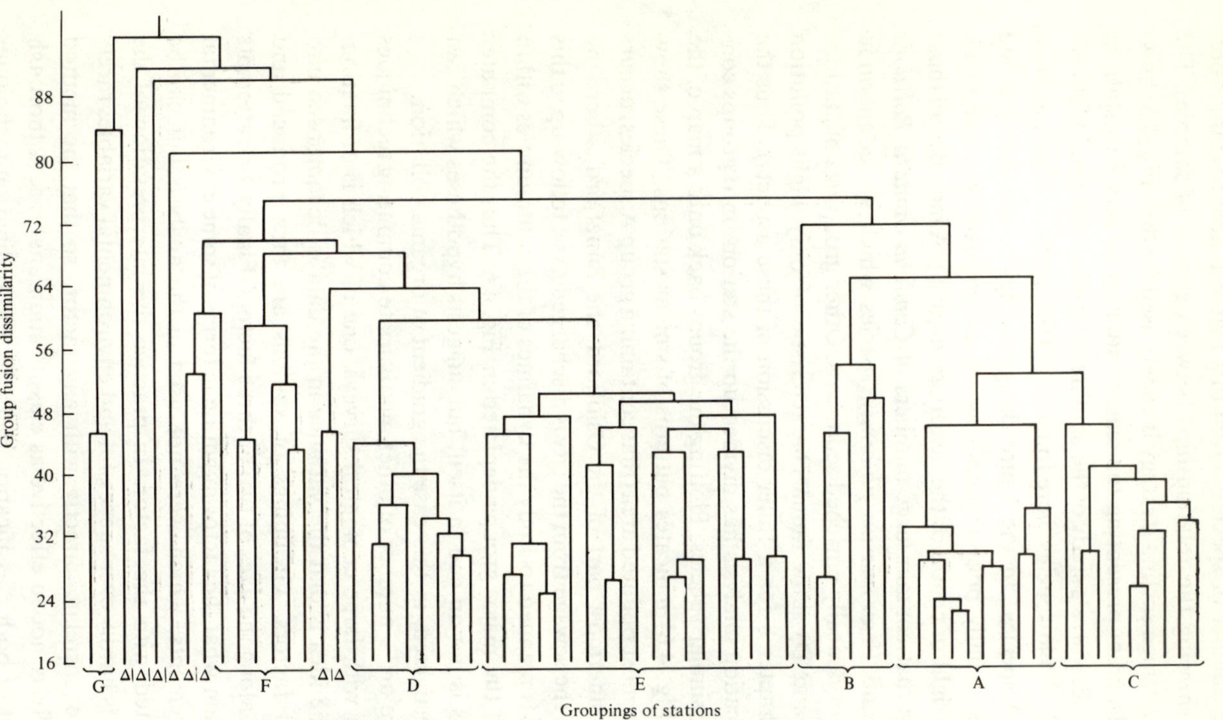

Fig. 4.5. Classification analysis of the Oslofjord based on 68 stations and 69 species. Based on Bray–Curtis dissimilarity index with group average sorting; data from F. B. Mirza. Species groupings are represented by letters A to F, isolated stations by triangles. Chaining (formation of a step-like structure where groups are not clearly dissimilar from each other) leads to progressively more 'misclassification' at the left-hand side.

Fig. 4.6. Plots on a map of Oslofjord of the major groupings (A to F) from the classification analysis shown in Fig. 4.5.

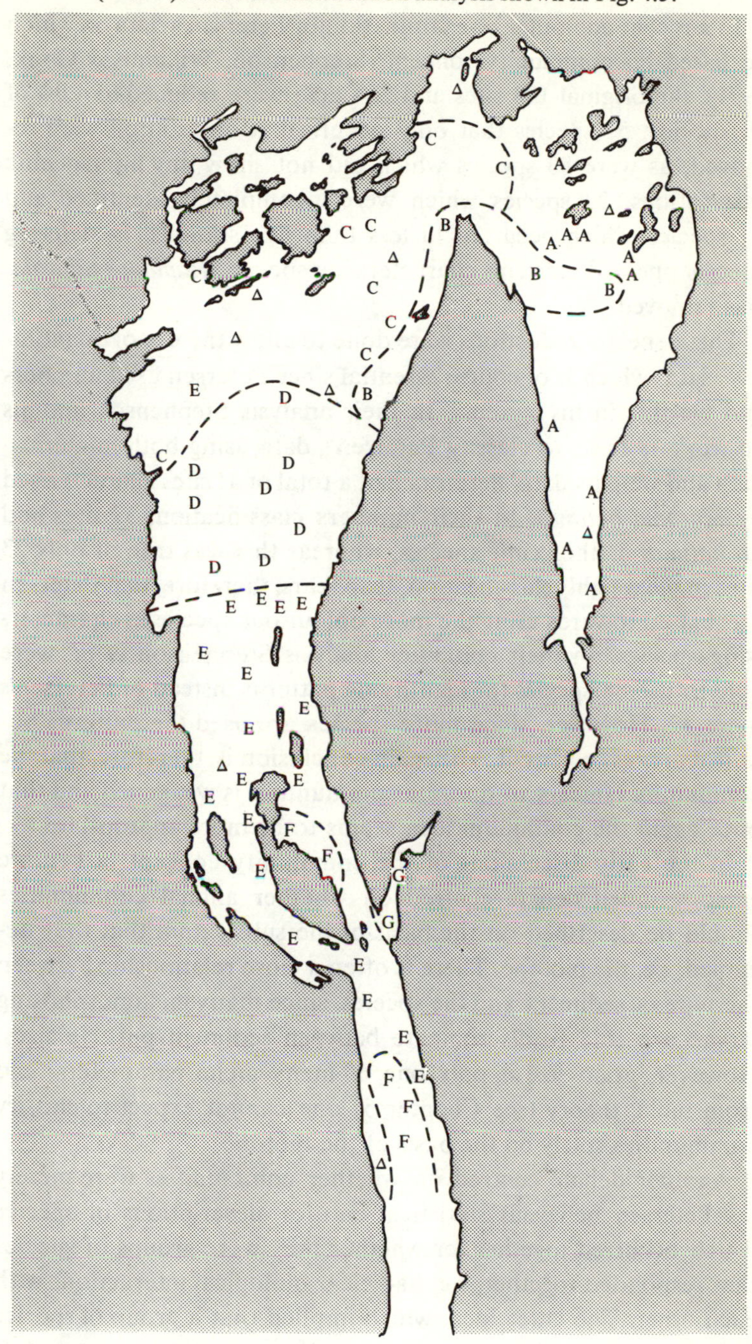

In a recent computer re-examination of Petersen's data on which he based his community concept (Stephenson, Williams & Cook, 1971), the original 193 sites and 264 taxa were reduced to a list of 88 species; 57 species that only occurred at one station were removed, as were 83 species which did not show any interspecific associations, 22 species which were incompletely identified and 13 species which occurred in less than five stations. In deleting species one of Petersen's characterising species, *Haploops tubicola*, was removed!

These species reductions were done to allow the use of quantitative data, which is of course essential since Petersen used numbers and weights in his systems. In their analysis Stephenson and his co-workers tried to classify Petersen's data using both numerical data and weights data. Petersen had a total of 41 sites characterised by *Macoma balthica*. In their numbers classification, 57 sites had *M. balthica* as the prime species, whereas this was true of only 23 sites in the weights classification. It appears, therefore, that Petersen was biased towards using the more ubiquitous species as characterising species. In their computer analysis Stephenson *et al.* were able to use 88 species for identifying patterns instead of Petersen's mere 12. However, since more species are used the patterns obtained are less clear-cut. The main conclusion is, therefore, that the community concept of discrete communities is weakened and that once again the continuum idea seems to be more appropriate.

In the early discussions of the community concept in benthic ecology much debate centred on whether animal communities should be described on the basis of the substratum that they occupied, i.e. the biotope. There is often a close relationship between the type of sediment and the species, since many metamorphosing larvae are able to discriminate between sediment particle sizes. However, grain size is only one of many niche parameters, and from niche theory (see Chapter 5) one cannot expect to classify communities solely on the basis of the biotope.

Another debate centred on whether communities were merely (as Petersen believed) statistical entities, descriptions of species which occurred together, or whether they were groups of species that responded together, i.e. had clear biological interactions with each other. The latter idea, which implied that a group of species had similar ecological requirements and hence acted as a cohesive unit, was termed a biocoenosis. Nowadays, we have come back to Petersen's idea of communities being merely helpful descriptive units, and not implying any rigidity in the patterns found.

Once species have been grouped then the next step is to try and see whether or not such groupings reflect similar preferences for a given set of environmental variables or are due to more complex factors such as biological interactions between species. The next chapter will deal with such problems.

5

The niche concept in benthic ecology

The most obvious way to unravel the factors influencing species in a community is to study a single species at a time. This autecological approach must involve both fieldwork to analyse the distribution of the animals in relation to measurable environmental variables, and laboratory experiments to find out the responses of the animals to these variables. By tradition the typical benthic ecological approach has been to concentrate on physico-chemical characters of the environment and usually to tolerance responses of organisms to these characters. Yet preference experiments give a more accurate picture of the likely responses of organisms to environmental factors. Furthermore, the limitation of investigations to predominantly physico-chemical factors has meant that benthic ecology has become divorced from other branches of ecology, where it has been clearly demonstrated that biological interactions give rise to many of the distribution patterns shown by species.

As an illustration of the type of approach used by benthic ecologists in investigating the factors controlling distribution patterns in the field, I will refer to my own study of a meiofaunal polychaete, *Protodriloides symbioticus* (Giard) (Gray, 1965, 1966*a*, *b*, *c*, *d*). I will not attempt to give a review of similar studies, but merely use this as an example of an approach.

5.1 Defining the niche of a species

Protodriloides symbioticus is 2 mm long and lives in large numbers between sand grains on intertidal beaches of Europe. The population that I studied in Wales occurred in dense patches on an apparently uniform beach. The problem was to unravel the factors that gave rise to the pattern. In laboratory experiments *P. symbioticus* tolerated salinity extremes of 18–55‰ for 12 hours and temperatures of −4 °C to +34 °C. Observations in the field over a 2-years period showed that natural variations in salinity and temperature were well within these limits. When given a temperature gradient of from 5 to 25 °C, *P. symbioticus* always chose 15 °C. Thus in response to too warm or too cold temperatures in the sand individuals would migrate to a preferred range. Yet counterbalanc-

Fig. 5.1. Responses of the polychaete *Protodriloides symbioticus* (Giard) to single environmental factors. Typical summer data are shown. The actual distribution is in response to the overriding factor of oxygen. (Stippled areas represent high densities of animals.)

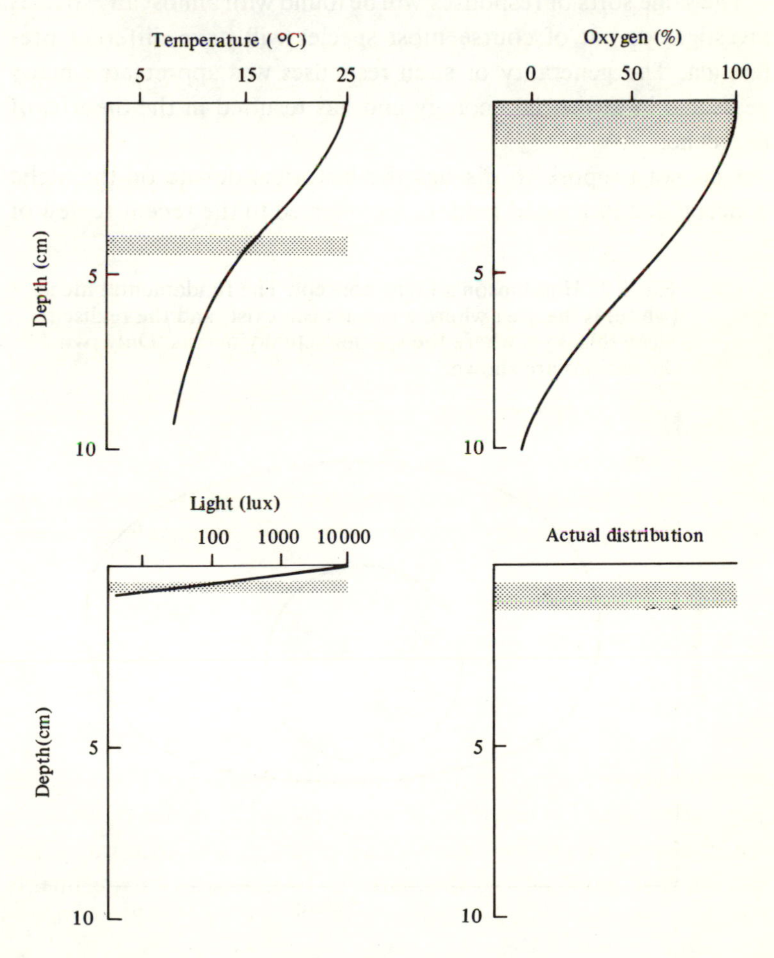

ing this response are the worm's reactions to light and oxygen. From laboratory experiments it was found that *P. symbioticus* has a preferred illumination range which under natural conditions would keep the species at a depth of 4–5 mm below the sediment surface. In oxygen gradient experiments, however, the organisms always preferred the area of maximum saturation, which on the beach is at the surface. This oxygen response overrides the temperature response.

Thus a combination of these responses explains the vertical distribution of the individuals in the sediment. Fig. 5.1 illustrates these responses diagrammatically for summer conditions.

The same sorts of responses will be found with almost any species investigated, but of course most species will have different preferenda. The generality of such responses was appreciated many years ago in terrestrial ecology and has resulted in the theories of the niche.

I do not propose to discuss the historical debate on the niche concept, and interested readers are referred to the recent review of

Fig. 5.2. Hutchinson's niche concept. The fundamental niche (white) is the area where a species can exist, and the realised niche (black) is where the species actually occurs. Only two dimensions are shown.

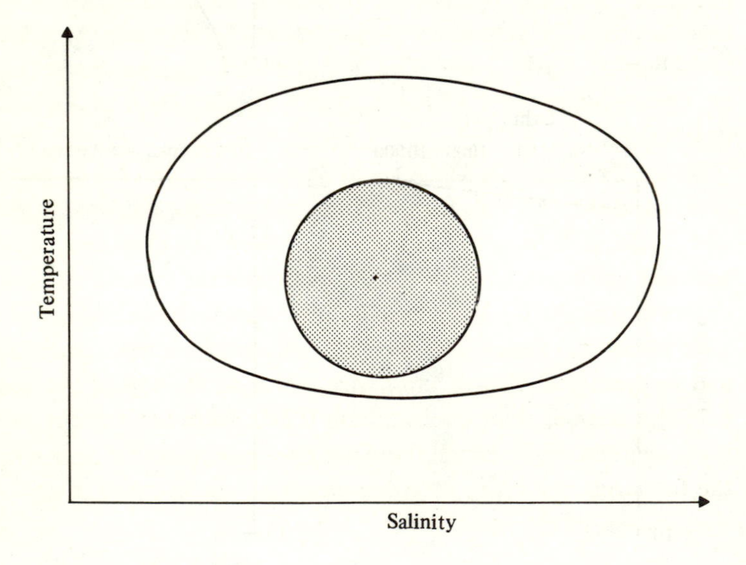

Vandermeer (1972). The most widely accepted view is that of Hutchinson. He divided the niche into two parts. The first of these is the fundamental niche, which contains the full range of environmental conditions under which a species *can* exist. No species in fact ever occurs over the full range of the fundamental niche, but is restricted to a part of the area. This is called the realised niche and is defined as that area where the species *does* exist. The constriction of the fundamental niche is due in part to preferenda for certain environmental conditions and in part to effects of biological interactions such as competition and predation which prevent a species from occupying the whole of its fundamental niche. Fig. 5.2 illustrates this idea, simplified here to the two dimensions of temperature and salinity. The actual number of dimensions that defines the niche for any species is the number of environmental variables that affects that species, and hence Hutchinson's idea has been called the *n*-dimensional niche hypothesis. The niche space can never be determined for a species since no one can ever be sure that he has measured all the appropriate dimensions. However, the hypothesis has proved extremely useful in the study of the factors affecting the distribution of species.

In the case of *P. symbioticus* the fundamental niche dimensions for temperature and salinity are − 4 °C to + 34 °C and 18–55‰ respectively. But the organism's preferred temperature of 15 °C together with the responses to light and oxygen greatly reduce the fundamental niche space. The actual realised niche, however, is an even smaller part of the potential fundamental niche so far described.

In the field the population was localised in patches in narrow areas at mid-tide level. It seemed probable that variations in grain size could influence the organisms if they could in fact choose between different sizes. Fig. 5.3 illustrates the results of such a choice experiment. It shows that the fundamental niche of *P. symbioticus* has dimensions not only of salinity, temperature, light and oxygen, but also of grain size, since at the top of the beach particles coarser than the optimum 0.2–0.3 mm are present and at the lower end of the beach there are finer particles. Yet even within the favoured size range *P. symbioticus* did not occur in a uniform distribution but in patches.

Fig. 5.3. Grain-size preference of *P. symbioticus*: (*a*) in laboratory experiments; (*b*) in the field, from two habitats in Anglesey, Wales (solid line) and Roscoff, France (broken line).

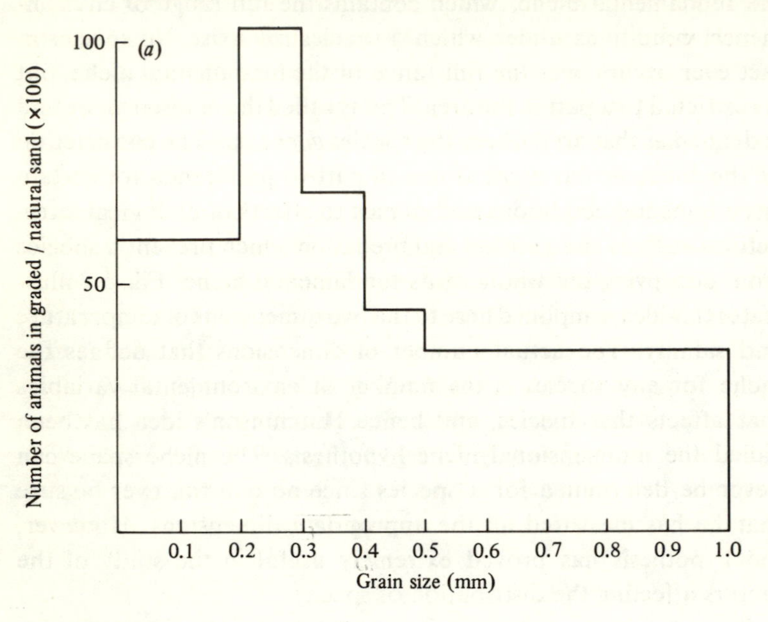

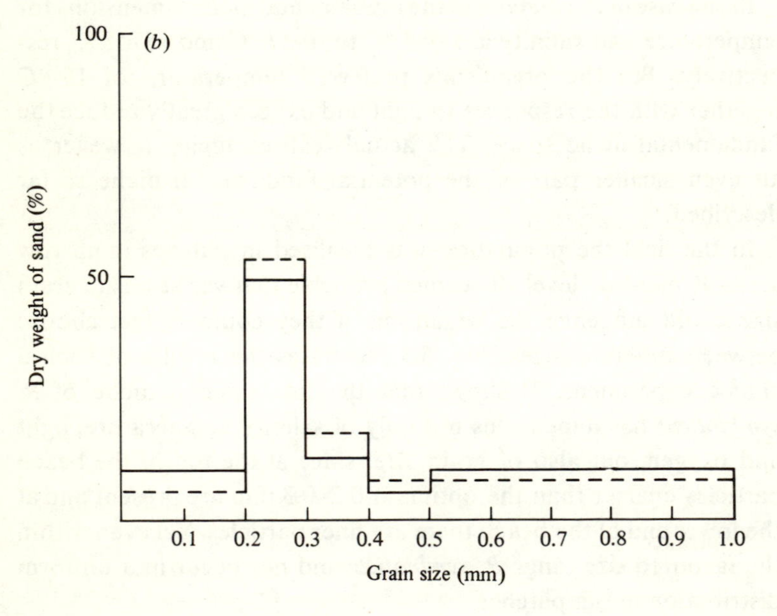

Sands from ten localities which had contained varying abundances of *P. symbioticus* in the field were, after removing the individuals, used in a multiple-choice experiment. Table 5.1 shows the results, which indicate that the order of preference corresponded to the abundances of the animals in the field samples. Certain sands had, therefore, an inherent attractive property. After a series of experiments I was able to show that this property was a favourable species of bacteria. *P. symbioticus* could distinguish between different species of bacteria and was localised in patches in response to the bacteria. Work on another meiofaunal species, done in collaboration with a bacteriologist, suggested that detection of the favoured species was by a tactile chemical response to the bacterial cell walls. *P. symbioticus* does not give off a chemical which attracts other individuals to a certain place. Subsequent to this work, a colleague found that *P. symbioticus* will not enter sands that have contained the gastrotrich *Turbanella hyalina*. The avoidance is a response to a chemical produced by *T. hyalina*, which thus restricts potential niche space for *P. symbioticus* by interference competition.

Table 5.1. *Preferences of* P. symbioticus *for ten natural beach sands in a multiple-choice laboratory experiment*

Rank of abundance in field sample	Rank in three separate experiments			Total rank in experiments
	A	B	C	A + B + C
1	1	1	7	1
2	7	2	1	2.5
3	3	9	5	5.5
4	10	5	3	7
5	5	8	9	8
6	2	4	4	2.5
7	4	7	2	4
8	9	6	10	10
9	8	10	6	9
10	6	3	8	5.5
Significance of correlation:	$P = 0.15$	0.15	0.15	0.04

The above experiments have shown that the realised niche of *P. symbioticus* is only a very small part of the original fundamental niche described by tolerance responses, and illustrate the importance of preference experiments and the experimental approach as tools in unravelling niche dimensions. Yet even though the investigations were fairly thorough, the potential impact of biological interactions was not studied, and the impact of predators especially may be very important.

In Chapter 10 I will discuss important new findings on the effects of competition and predation on benthic species. Investigations of such factors with meiofaunal species is difficult because of their small size; also, since many are entirely soft-bodied, they do not leave traceable remains in the guts of their predators. One approach to overcome this difficulty is to use a technique developed in terrestrial biology and applied to feeding studies of freshwater triclad turbellarians. The prey organism is freeze-dried and injected in a concentrated suspension into the hind-leg muscle of a rabbit. With a repeated series of injections over three to four weeks the rabbit builds up antibodies to the prey organism. A blood sample is then taken from the ear and, after the red blood cells have been removed by clotting, the blood serum containing the antibodies can be frozen for storage. The thawed antibody solution is placed in a central well cut in an agar-gel plate and gut contents from suspected predators placed in wells cut round the periphery of the plate. If a predator has eaten the prey, antigens from the prey will be present in its gut contents. These antigens, as they diffuse through the gel, will meet the antibodies diffusing out from the central well and the two will combine to form a white precipitate. The presence of a precipitate thus indicates that a particular predator has eaten the prey. In a study of meiofaunal feeding on a beach of England I used this method to screen 12 species of meiofauna against one species of bacterium (the favoured one from the *P. symbioticus* study) and found that at most five meiofaunal species living in the same area were utilising the bacterium. The studies on freshwater triclads described above have unravelled a great deal of new information on feeding types and the technique should be a great boon to our understanding of feeding strategies in many benthic species. Very recently an important paper has appeared using the above technique on a marine benthic community and it clearly illustrates the effectiveness of the method in studying predator–prey relationships (Feller *et al.*, 1979).

In general, though, there are very few data available on niche dimensions and thus few investigations of niche overlap and competitive interactions have been attempted in benthic ecology. A notable exception is provided by the fascinating studies of Tom Fenchel (1975) and his co-workers on hydrobiid snails on sandy and muddy beaches in Denmark.

5.2 The uniqueness of the niche and character displacement

The niche of a species is supposedly unique to that species. So the commonest pattern that can be expected in nature is that closely related species are separated along one or more niche dimensions which are important to the species. In the Limfjord in North Jutland three species of the genus *Hydrobia* are very common, and their distribution in the field suggests that salinity is an important niche dimension. The results of preference experiments in the laboratory and those from field investigations are summarised in Table 5.2. Clearly, there is considerable overlap in the salinity ranges where these species are found in the field, and their distribution pattern cannot be exclusively related to salinity preferences. *H. neglecta* should, in terms of salinity preferences, be the commonest species, since most of the habitats investigated had the preferred salinity for this species. *H. ulvae* occurred predominantly in the deeper, more saline areas, but its distribution extended right into the fjords, where it excluded *H. neglecta*. Fenchel suggests that *H. ulvae* is a superior competitor to *H. neglecta*. Yet, as the salinity

Table 5.2. *Salinity preferences in laboratory experiments, and field salinities from localities in Limfjord, Denmark, for three species of* Hydrobia

	H. ventrosa	*H. neglecta*	*H. ulvae*
Laboratory preference (‰)	20	25	30
Field range (‰)	6–20	10–24	10–33

Data from Fenchel (1975).

is reduced, *H. ventrosa* becomes a superior competitor to *H. ulvae*. *H. neglecta*, therefore, is confined to areas with low population densities of the other two species. Where the changeover occurs from one species dominating to another species dominating depends on the intensity of the competition between the species, the relative competitive abilities at different salinities and the rates of dispersal.

In his studies, Fenchel has concentrated on competition between *H. ventrosa* and *H. ulvae*. Niche theory states that no two species can coexist on the same limiting resource; one or other of the species must win control of the resource, while the loser either becomes extinct or, over evolutionary time, develops specialisations so that the two species no longer compete. The populations of *Hydrobia*

Fig. 5.4. Character displacement in two *Hydrobia* species from Denmark. (*a*) Shell lengths from localities where species occur alone (i.e. are allopatric); (*b*) shell lengths from localities where species coexist (i.e. are sympatric). (From Fenchel, 1975.)

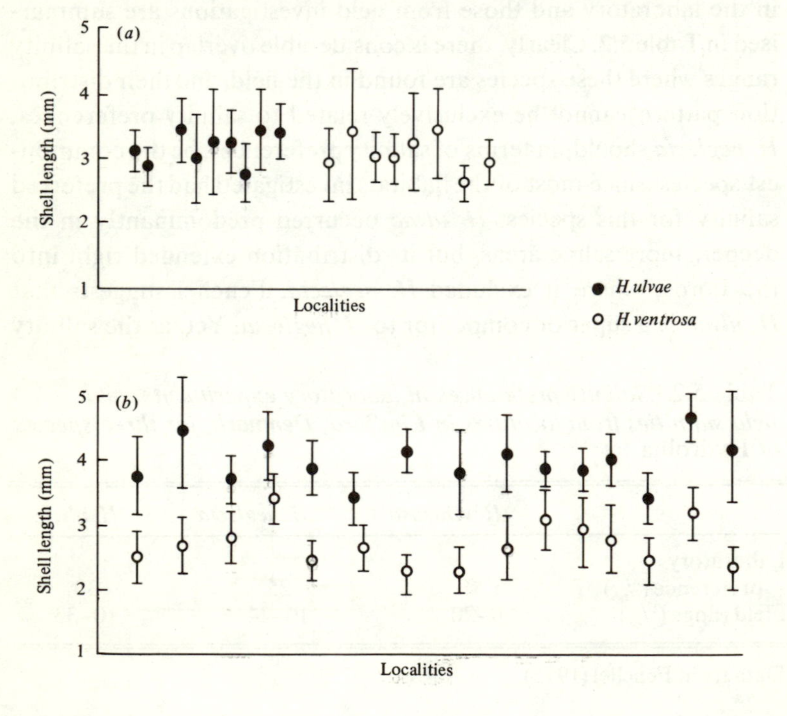

occurring in the Limfjord have very high population densities and are almost certainly food-limited. Studies of their feeding habits indicate that all three species are indiscriminate feeders, ingesting sediment and utilising micro-organisms. They all appear to subsist on the same food items. Yet the size range of particles ingested is dependent on shell length. Thus, if two species have different shell lengths, they can coexist because they are not utilising the same food resource. When *H. ventrosa* and *H. ulvae* live allopatrically they have almost identical size ranges (Fig. 5.4*a*), but it is remarkable that when they occur sympatrically their size ranges differ (Fig. 5.4*b*). This type of mechanism whereby two species avoid competing for a

Fig. 5.5. Ingestion of food particles in two *Hydrobia* species. (*a*) and (*b*) Allopatric pattern; (*c*) sympatric pattern. (From Fenchel, 1975.)

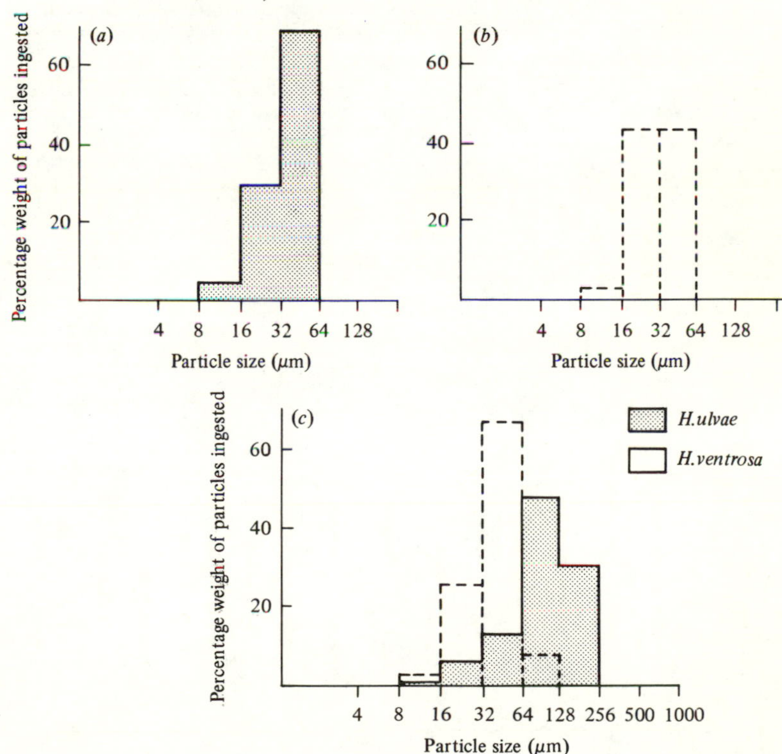

limiting resource is called character displacement. Curiously enough Hutchinson had predicted that if food was the limiting resource the important dimensions of the mouth-parts of two related species should differ by 1.3 units to avoid competition. Fenchel's data show that on a $\log_2$ scale, the appropriate scale, Hutchinson's prediction is borne out (Fig. 5.5). These elegant studies indicate the opportunities awaiting persons willing to apply simple ecological theory to marine benthic species.

6

Diversity

One of England's most illustrious nineteenth-century naturalists, Edward Forbes, predicted in 1843 that no animal life would be found below a depth of 550 m since there was no light and the pressure was too great. Forbes had overlooked the work of the Rosses (Sir John and Sir James Clark), who as early as 1817 had obtained many living animals from depths of 1800 m in Baffin Bay, Canada. In 1869 the Sars (M. and G.O.) trawled up a variety of marine life from depths greater than 550 m off the Norwegian Lofoten Islands. Forbes views, however, were influential, and there was still no general acceptance that life existed at great depths. Thus, when the Royal Society set the aims for the *Challenger* expedition, one of the problems to be tackled was the distribution of organic life at all depths of the ocean and on the sea-bed. This expedition, which lasted from 1872 to 1876, was the first truly scientific oceanographic expedition; it obtained, in all, 133 dredge samples from the deep sea, which showed conclusively that life does exist at great depths. The number of animals obtained was relatively small, but most were new to science. For almost the next 100 years the pattern was repeated, with expeditions finding many new species of deep-sea animals, but each species being represented by relatively few individuals. Initially this pattern was thought to be an artifact resulting from the fact that the trawls and dredges were not closed, for it seemed likely that many animals were winnowed out on the way up from the great depths. In the 1950s and 1960s better gear was available and it became clear that the earlier expeditions had certainly lost material and that the density of animals was higher than they had found; but equally the pattern of a high number of species for a relatively small number of individuals was consistent in almost all taxonomic groups. The reason for the low density of organisms on the deep-sea floor is clearly the low amount of food available, either as settling organic matter or produced *in situ*. The settling organic material is broken down on its passage downwards and relatively little nutrient remains by the time the particles reach the sea-bed. Similarly, since there is no light there is no primary production on the sea-bed, and bacterial production must be limited since the organic substrate is limited.

This last fact has been confirmed in a rather curious way. The Wood's Hole Oceanographic Institute lost its deepsea submersible *Alvin* overboard from a research vessel. *Alvin* was at the time ready to make a dive. The crew's sandwiches were on a table and the hatch cover was open. Some 10 months later *Alvin* was recovered from a depth of 1540 m and the sandwiches were found to be in perfect condition, suggesting that bacterial action is very slow at the bottom of the deep sea! Subsequently, more sophisticated experiments have shown that bacterial processes are roughly 10–100 times slower at 4000 m than in controls kept at 3 °C in total darkness in the laboratory.

The consistent pattern, then, in the deep-sea benthic data is that, compared with samples taken from shallow depths, the number of species is remarkably high in relation to the number of individuals found. A similar pattern is found in terrestrial habitats when one compares the rich tropical with boreal fauna. If these patterns are to be investigated in a scientific way, then we must attempt to quantify the richness in species of the deep sea and tropics. This is usually done by calculating a diversity index.

To an ecologist diversity intuitively means more than just the number of species. For example, if one community has 50 individuals of each of two species A and B whereas another community has 99 individuals of species A and 1 individual of species B, then the first community is the more diverse. Thus a diversity index must take into account not only the number of species but also the number of individuals per species. If fact, there are a large number of diversity indices and I do not propose to consider them here; Hurlbert (1971) has given a critical and extensive review. Instead, I shall describe the two indexes most commonly applied to the fauna of marine sediments, namely the Shannon–Wiener information statistic and the rarefaction method.

6.1 Measuring diversity

6.1.1 *The Shannon–Wiener information statistic*

The Shannon–Wiener information statistic relates cybernetic theory to ecology. The organisation of a community can be represented by the number of species and the number of individuals per species. This organisation is acted on by environmental factors to produce, perhaps, a different organisation in the future. The cybernetic analogy comes in by equating the organisation to an information channel running from the present state into the future, with the width of the channel being a measure of the organisation (Margalef, 1968). More explicitly, the formula for calculating diversity is:

$$\sum_{i=1}^{s} pi \log_2 pi,$$

where $pi = ni/N$ (ni being the number of individuals of the ith species and N the total number of individuals) and s is the total number of species.

This may seem a little complicated and it is hard for a biologist to envisage what the index is actually measuring. So let us take a simple example. If there are four species A, B, C and D, each with the same number of individuals per species (say two), what the index measures is how many binary decisions ($\log_2$) are necessary to decide whether a new individual belongs to the above four species or not. This can be illustrated as

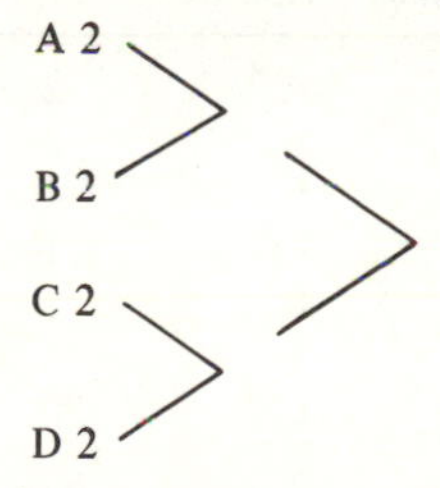

Firstly we decide between A–B and C–D, and then between A and B or C and D; thus two binary decisions are necessary and application of the above equation will give a value of 2. If we add species E, F, G and H, each also having two individuals per species, then we require three binary decisions and the index works out at 3. Of course natural communities are much more complex, with many species and with unequal proportions of individuals per species. In general, as the number of species increases so does the diversity index. But the diversity index will also increase as the proportion of individuals per species becomes more constant. Table 6.1 illustrates some simple hypothetical examples.

The diversity index thus measures two things: species richness and species evenness. The latter can be separated from the index by dividing the observed diversity value by the maximum possible value which would be obtained if each individual belonged to a different species. Evenness (J) is thus defined as

$$J = H'/H'_{max},$$

where H' is diversity and $H'_{max} = \log_2 s$. Unfortunately the value of the diversity index is often reported merely as H' and the evenness component is not given. One is then not sure whether the value of any change in diversity found is due to an increased number of species or a more even distribution of individuals per species. Table 6.1 also shows values of evenness J.

Table 6.1. *Theoretical examples of diversity (H') and evenness (J) showing effects of adding rare individuals and of increasing dominance*

Species	Number of individuals	Species	Number of individuals	Species	Number of individuals
A	2	A	2	A	2
B	3	B	3	B	3
C	4	C	4	C	4
D	1	D	1	D	1
E	1	E	1	E	1
F	6	F	6	F	6
G	10	G	10	G	50
7	27	H	1	7	67
		I	1		
		9	29		
$H' = 2.4036$		$H' = 2.6688$		$H' = 1.4025$	
$J = 0.8562$		$J = 0.8419$		$J = 0.4996$	

If one plots for typical benthic communities diversity (H') against $\log_2 s$ and diversity against evenness (J), one can establish whether the diversity index is more responsive to an increase in the number of species ($\log_2 s$) or to an increasing evenness in the distribution of individuals among species. Fig. 6.1 shows some data for benthic nematode communities in the North Sea. Diversity is usually poorly correlated with $\log_2 s$ and has a better correlation with J. Thus addition of rare species to the community has little effect on diversity whereas changing dominance has a larger effect. Similar plots can be obtained for most subtidal macrobenthos and meiobenthos. Not all communities however, show such trends. Bird communities show a higher correlation of H' with $\log_2 s$, indicating that addition of rare species is relatively more important than a change in dominance pattern. An explanation for this may lie in the fact that bird species are largely territorial and that as a consequence dominance patterns remain fairly constant.

Fig. 6.1. Relationship between diversity (H) and evenness (J) for nematodes in southern North Sea sediments. (Data from Heip & Decraemer, 1974.)

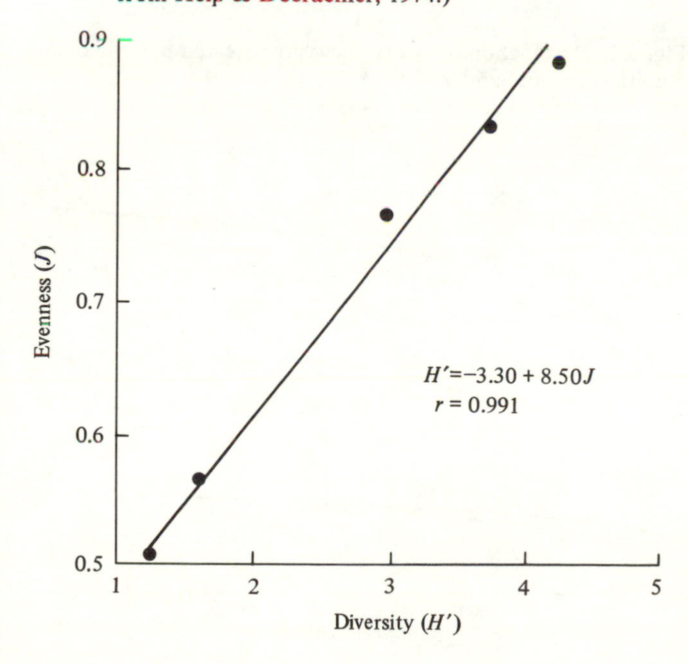

6.1.2 *The rarefaction method*

The other method of measuring diversity which has been widely used with marine benthic data is the rarefaction method of Sanders (1968). This is a graphical method of expressing diversity (Fig. 6.2), steep curves showing high diversity and shallow ones low diversity. The calculation begins with the total number of species and individuals and following a set method calculates how many species would be found in a smaller sample. Hurlbert (1971) has suggested that the method as described by Sanders overestimates some fractions, and he has made a correction.

Having discussed the methods for measuring diversity, let us return to the important question of the patterns of diversity in benthic communities and, in particular, why diversity is so high in the deep sea.

6.2 The diversity of benthic communities

Sanders (1968), in his paper describing the rarefaction method, stimulated a debate that caught the imagination of many workers and resulted in a whole new direction in ecological re-

Fig. 6.2. Rarefaction diversity curves of selected benthic communities. (From Sanders 1968.)

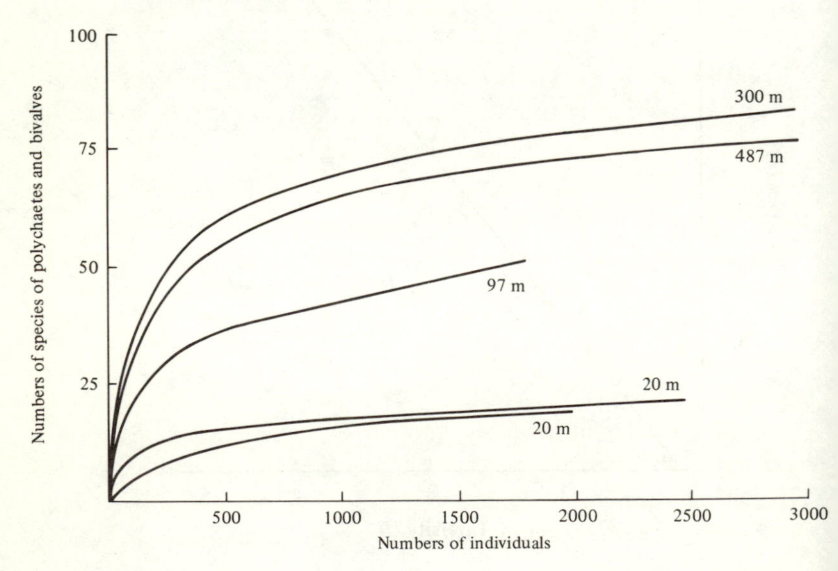

search. Working at the Wood's Hole Oceanographic Institute, Sanders had been studying the benthos of the deep sea for many years. He had amassed data from a wide variety of depths and geographical regions and was struck by two things. Firstly, the deep sea had in fact an amazingly high complement of species, although the number of individuals per square metre was low. Since the total number of individuals is low and the number of species high, diversity is high however one measures it. Secondly, Sanders showed that, just as on land, the tropics had a higher diversity than boreal regions. A possible explanation is that natural selection has had a longer time to act in the tropics than it has in the ice-age-prone polar and boreal regions. The processes leading to the high tropical diversity are, however, still under dispute. Another theory is that competition there is intense, which means that niches are smaller and so there are more species per unit area. A third theory is that there are more predators in the tropics, which keeps the abundance of prey species low, prevents competition and thereby allows more species to coexist. The competition and predation theories appear to be, therefore mutually exclusive, and as will be shown later are at the root of the deep-sea diversity debate.

Sanders' rarefaction curves (Fig. 6.2) indicate that deep-sea diversity is very much higher than that of shallower areas. He explained this high diversity by the stability-time hypothesis. He postulated that near one environmental extreme, in the high intertidal, the fauna is subjected to environmental factors that fluctuate in an unpredictable manner, and as many species are not able to tolerate these unpredictable fluctuations the species complement is low. At one point in time species A may be dominant, but competitive exclusion does not occur because before this can happen the environment changes, giving a competitive advantage to species B. Sanders suggests that this results in species in intertidal areas having broad overlapping niches. However competition and predation effects also operate, and these lead to large fluctuations in population sizes and low diversity. This part of Sanders' argument has been misunderstood and wrongly quoted. Sanders regards species as adapting to the environment and not to each other – hence he calls this the physically controlled habitat. However he does *not* say, as he has been quoted as saying, that there are no biological interactions here. In fact, biological interactions of competition and predation may be very severe. The important point is that niche specialisations do not occur since the environment is constantly fluctuating.

By contrast the deep sea is an extremely constant environment, with no light and almost no changes in temperature, salinity or oxygen from month to month and year to year. Furthermore, it has remained constant for a very long time-period (probably thousands of years) compared with the glaciated boreal and polar regions. This constant environment over evolutionary time-scales has enabled the species to adapt to each other rather than needing to adapt to the rigours of the environment, as is the case in the intertidal area. The deep-sea species at one time competed for the most important limiting resource: food. Since the available food reaching the deep-sea bottom from the surface is small, densities of animals per square metre are low. Over evolutionary time they have become what Sanders calls 'biologically accommodated' to each other and now have narrow non-overlapping niches. The main point of Sanders' argument, then, is that competition is the causal mechanism that, over evolutionary time, has led to high deepsea diversity. The problem is that this must remain a hypothesis since it cannot really be tested.

Following the publication of Sanders' stimulating paper, Dayton & Hessler (1972) from the Scripps Institute of Oceanography at the University of California suggested that rather than competition as the causal mechanism of high deep-sea diversity, predation was the key. They argued that there was no evidence that deep-sea species were more specialised than their shallow-water counterparts, as would be necessary in a competitive situation. Dayton and Hessler managed to do some experiments *in situ* in the deep sea. They lowered bait (dead fish) and arranged a camera with flash so that any species coming to the bait could be photographed over various time-periods. Surprisingly they found that high abundances of fish, amphipods, isopods and ophiuroids were attracted to the bait within a few hours, in an area where food was supposedly very scarce. Dayton and Hessler termed these organisms 'croppers' rather than predators since they surmised that their effect on the benthic sediment of living organisms may not be a direct predator–prey response but rather unselective feeding or even chance destruction by the cropper without being ingested. It is these croppers, they suggested, that account for the high deepsea diversity, by holding benthic population densities below the level where competitive exclusion can occur. (This is an analogous argument to that that has predators being responsible for high tropical diversity.) Since food is argued as being the limiting resource in the deep sea that is competed for, all the deep-sea species with non-overlapping niches must have specialised in utilising the food in different ways in order to avoid competing with each other. Dayton and Hessler also reviewed all the available data on feeding specialisations in deep sea infaunal species and suggested that rather than their being specialised, most were generalists eating anything that was available, which is in keeping with the predictions of the cropping argument.

Back came Sanders with a counter-argument, with Grassle as co-author (Grassle & Sanders, 1973). They showed by the size-frequency distributions of known deep-sea species that there was a high proportion of older individuals. If cropping was indeed general in the deep sea, one would expect that there would be a superabundance of young stages, since in terrestrial habitats prey species adapt to high predation pressure by producing more young, thus ensuring that a sufficient number of individuals survives to reproduce. But size-frequency analyses of the deep-sea species that have been studied show that there are relatively few individuals in the smallest classes. Furthermore, Grassle & Sanders argued that feeding specialisations need not be to different food items (such as mechanisms for dealing with different sized particles) but could be biochemical in nature, such as those specialisations in digestive enzymes that are known to allow up to four species of polychaete to coexist on apparently the same food source. Deep-sea species may well show such specialisations, but no one has as yet tested this suggestion.

There the argument largely rests, as two apparently opposing hypotheses. Jumars (1975), one of Hessler's students, took the obvious next step: he attempted to measure another possible source of niche specialisation, spatial pattern, that would result if there were competition for space between deep-sea species. The argument here is that species may show more intense patchiness in the deep sea than in shallow areas, since deep-sea species are supposedly more specialised. He took large box-corer samples of deep-sea sediments and divided up the samples into many small subsamples. He found that most species were, in fact, randomly distributed and there was no direct evidence that patchiness was more intense in deep-sea species. The problem is that since the food supply to the deep sea is so low and therefore densities are low, much greater areas must be sampled in the deep sea as compared to shallow areas in order to determine distribution patterns. So it is very difficult to make direct comparisons. On cannot expect to find the same patterns in the deep sea and shallow areas.

The latest development in this fascinating topic has come from Huston (1979), who has treated diversity patterns in general and not specifically deep-sea diversity. Huston noticed a number of anomalies in the arguments as to why diversity was high in some areas and lower in others. For example, none of the current theories on diversity can explain the low-diversity communities which occur in highly predictable environments (coastal redwoods and freshwater marshes) or the high-diversity communities in unpredictable environments (Sonoran desert and some marine communities).

Huston's argument is that there is a population level at which a species begins to compete with other species. This he calls competitive equilibrium and it varies from species pair to species pair. If there are many species in a community all with low growth rates, then competitive equilibrium will not be reached for a long time and diversity remains high over the time-period in question. If one species grows fast, however, then it can rapidly reach competitive equilibrium and begin to exclude the other species. Here diversity falls. So one important variable determining diversity values is the relative growth rates of species within a community.

If the species which potentially can outcompete the other species in the above case is removed, either by predators or by an environmental factor, then diversity can still be kept high since competitive exclusion does not occur. So the frequency of population reduction is another important variable affecting diversity. Fig. 6.3 illustrates these ideas. Where growth rates of competitors are low, diversity is low at low frequencies of disturbance, provided that the time between successive disturbances is long enough for species to approach competitive equilibrium. Moving up the growth rate axis, diversity is maximised when the frequency of disturbance just prevents competitive equilibrium being reached. If growth rates of all competitors are high, then diversity will be low because at low frequencies of population reduction there is a rapid approach to competitive equilibrium. Along the other axis, at a low frequency of disturbance diversity rises rapidly with a small increase in growth rate. But if frequency of disturbance rises when there are low growth rates, diversity becomes low since some species are eliminated. With high frequencies of disturbance and high growth, a higher diversity will be achieved because populations are able to recover from the disturbance. To me, Huston's argument is very convincing. The suggestions that predation is just one of a number of potentially disturbing agents and that the effects of predation are not necessarily predic-

Fig. 6.3. Huston's model of relationship between disturbance, growth rate and species diversity of a community. Contours are of equal diversity values with highest values innermost. (From Huston, 1979.)

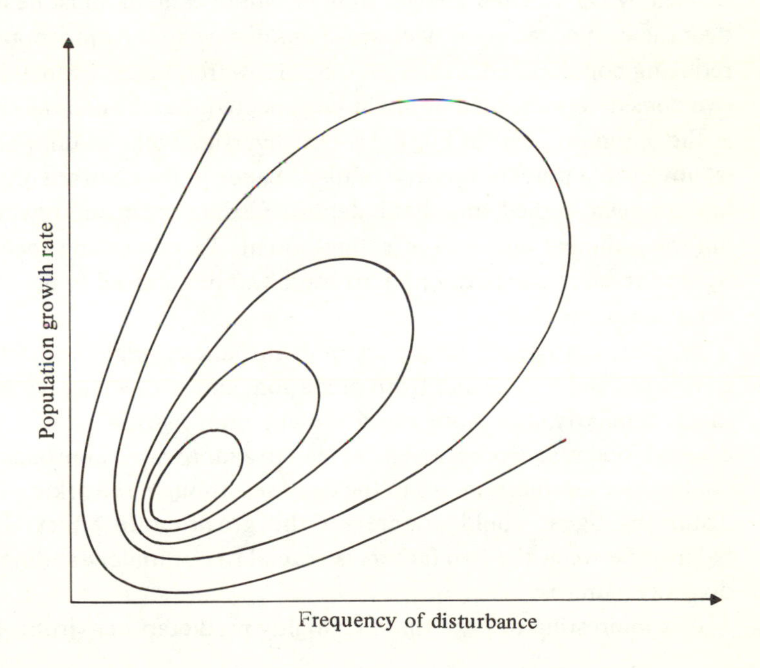

table are important ones. The so-called 'predation theory' will have to be revised. In a talk given some six months before Huston's article appeared I came to similar conclusions in trying to explain how benthic species adapt to pollution (Gray 1979*b*; and see Chapter 8). I believe that disturbance is the key factor operating in many pollution incidents and that adaptive strategies are to disturbance, not tolerance of chemicals. Huston's ideas are elegant in their simplicity and are much more comprehensive than my own!

The applications of Huston's arguments to the deep-sea controversy are these. Growth rates in the deep sea are probably exceedingly slow. Living specimens of the bivalve *Tindaria callistiformis*, which measures only 8.4 mm, were obtained from 3800 m depth and dated using radium (^{228}Ra) dating. *Tindaria* does not reproduce until it is around 50–60 years old and it can survive until over 100! If such properties are general, then growth rates are low. However, the important point as regards Huston's argument is *relative* growth rates and the rate at which populations approach competitive equilibrium. On this there is no information. It seems probable that disturbance frequency is also low. The chances of falling carcasses, as used by Dayton and Hessler in their baited camera, must be low over the whole area of the deep sea. Thus disturbance from croppers reducing population densities must be of low frequency. From these two considerations, then, it might be expected that the deep sea lies at the optimum point of Fig. 6.3 where diversity is maximum. There is, however, a potential source of disturbance in the deep sea which has not been studied, and that is deposit-feeding organisms reworking the sediment and thus affecting potentially competing species by interference competition. This topic will be referred to in more detail in Chapter 10.

In shallow areas the frequency of disturbance is high, both from environmental factors and from predation, and so diversity is lower there. Similarly, the more rapid growth rates associated with increased food resources also tend to lead to a more rapid approach to competitive equilibrium. As in the deep sea, though, reworking as a disturbing agent could counteract the growth rate effect. The balance between the two factors is crucial to our understanding of diversity patterns.

It is interesting to note that in a highly predictable environment with no environmental disturbance Huston's model predicts low diversity, whereas the stability-time hypothesis predicts maximal diversity. This point was raised earlier by a geneticist, Campbell, who could not accept that speciation would be highest in a highly predictable environment. He would endorse Huston's argument. Huston's model also predicts the low diversity that is found in stable areas with high amounts of organic matter (eutrophic areas), since growth rates are high as is disturbance by the continual rain of organic matter.

Before leaving deep-sea diversity, one other aspect, which has been neglected, should be considered, and that is that diversity does not in fact increase with depth. There appears to be a depth zone between 2000 and 3000 m, where diversity is highest. The deepest parts of the deep sea go down to over 10 000 m, and here diversity is lower. The very deepest parts are, however, in trenches and it could be argued that here occasional turbidity currents rolling down the slopes give a more unstable environment than in the flat plains, thus leading to lower diversity. But the average depths of the deep sea are around 5000–6000 m and here diversity values are lower than on the slopes. A possible explanation is that diversity varies not only with time, environmental predictability, competition and predation, but also with the structural heterogeneity of the habitat. In sedimentary habitats structural heterogeneity can be measured by particle size and sorting coefficients. Where coarse sediments occur

Table 6.2. *Heterogeneity of selected sediments from the deep-sea slopes*

Station	Depth (m)	Mean grain size (phi)	S.D.	Diversity
Sl3	300	3.50	1.51	61
Dl	487	4.08	1.62	56
Fl	1500	6.81	2.09	58
Gl	2086	6.67	2.94	55
GH	2500	8.32	2.36	52[a]

Data from Sanders, Hessler & Hampson (1965). Diversity data from Sanders (1968), assessed as number of species per 500 individuals read from rarefaction curves.

[a] Extrapolated rarefaction curve to 500 individuals.

in nature, wave and current action are high, giving a harsh environment and low diversity. Fine muds and clays represent a stable wave and current regime, but are structurally homogeneous. A heterogeneous sediment of varying particles sizes (poorly sorted), on the other hand, can be expected to give more structural heterogeneity and potential niche space and therefore higher diversity. The fall in deep-sea diversity below 2000–3000 m is probably due to a reduction in structural heterogeneity as reflected by finer sediments (Table 6.2).

Diversity then, as this chapter has shown, is an ecological property of a community and should be used in a comparative way. A diversity value in itself means little. Attempts to give legislative values to diversity indices in relation to effects of pollution seem to me misguided. I will return to this aspect in Chapter 8, when considering effects of pollution on benthic communities.

7

Stability

When considering the stability of benthic communities one usually thinks of the patterns of numerical or biomass dominance over time. But what sort of time-scale is important in this context? Are the patterns repeatable over short periods of less than a year or do they run over longer periods? Perhaps it is surprising that there are remarkably few long-term data-sets that can be used to answer such questions. Most benthic investigations have been geared to studies done over the three years of a PhD thesis or are connected with surveys before and after effluent discharges, where the objective is not necessarily to study the long-term patterns. Since time-scales are obviously a crucial aspect in considerations of stability, let us look at the patterns that are obtained over various time-scales.

7.1 Patterns of less than one year

In his classical studies of the benthos of Long Island Sound, New York, Sanders (1956) found that over a 2-year period only two species constituted more than 10% of the total number of individuals (except for one month when the cephalocarid *Hutchinsoniella* reached 12%), although there were a total of 98 species in the community. Fig. 7.1 shows the data. Clearly, there is an oscillatory pattern within each year, and when *Nucula* is dominant *Nephthys* is not, and vice versa. This pattern of oscillating dominance is actually quite common.

One of the most interesting examples of such cyclical oscillation patterns that has been investigated is from a study in Barnstaple Harbor, Maine, by Eric Mills (1969). Mills found the pattern shown in Fig. 7.2. In the winter the sediment is dominated by the mud-snail *Nassarius obsoletus*, which occurs in large numbers and feeds on the detritus in the sediment. In spring a few individuals of the tube-building amphipod *Ampelisca abdita* become established and since

Ampelisca (like all amphipods) is a brooder, a rapid build-up of the population occurs. The tubes of *Ampelisca* hinder the feeding of *Nassarius* and so the latter population declines, outcompeted for space by *Ampelisca*. The tubes increase the spatial heterogeneity of the habitat and other species are able to coexist with *Ampelisca*,

Fig. 7.1. Changes in dominance of infauna from a sandy habitat, Long Island Sound, New York. ●—●, *Nucula proxima*; Δ—Δ, *Nephthys incisa*. (From Sanders, 1956.)

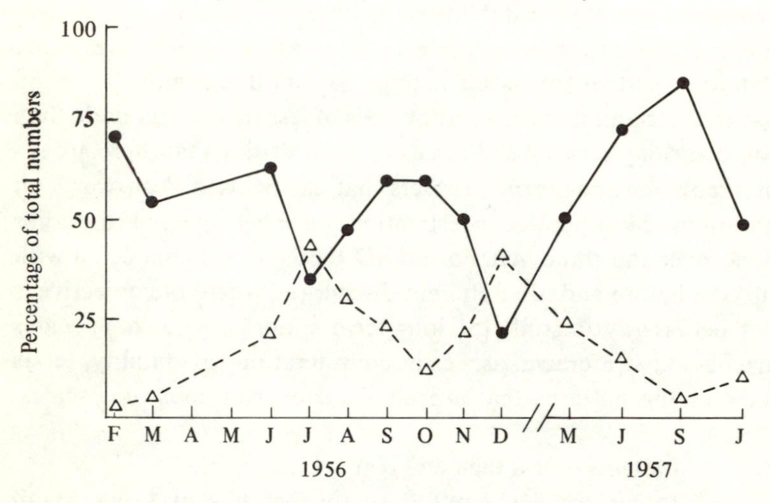

Fig. 7.2. Schematic representation of changes in dominance patterns of an intertidal mudflat, Barnstaple Harbor, USA. Solid line, *Ampelisca*; broken line, *Nassarius*. (Data from Mills, 1969.)

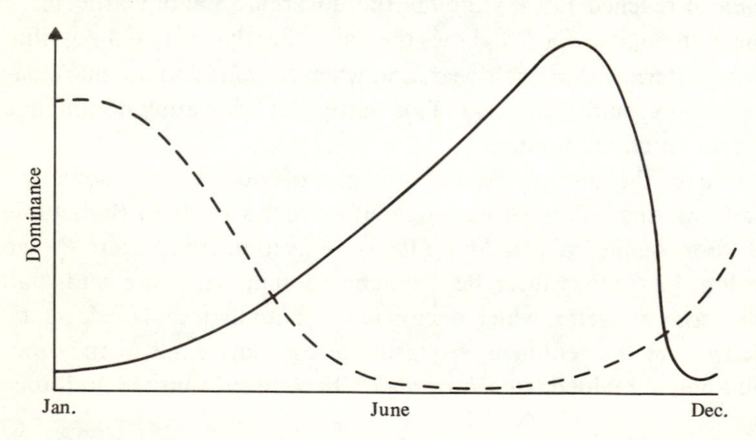

subsequently increasing species richness. Then in the autumn, storms begin to dislodge the *Ampelisca* tubes, which are rolled up like a carpet, and a rapid decline in the population results. The sediment is then available for recolonisation by *Nassarius*, and so the cyclical pattern continues. This pattern of one species making the habitat unsuitable for another species is a rather interesting property of sediment-living species and does not occur to the same extent on rocky shores. There are many known cases of it and I have reviewed the data (1974). One point that should be borne in mind is that the first example given here was concerned largely with a single station, whereas Mills' data concern a much larger area, in this case many hundreds of square metres. I shall return to this point later.

One of the simplest and yet most effective ways of demonstrating that competition can be responsible for changes such as those shown above is by manipulating the natural populations. Pioneering research in this area was done by Sally Woodin (1974) at the University of Washington. She excluded the settlement of a tube-building polychaete by means of a cage placed over the sediment on an intertidal mud flat, and was able to show that this resulted in a dramatic increase in the abundance of a burrowing polychaete (Table 7.1). Within a period of only four months she was able to demonstrate the importance of competition for space, which would have taken decades by traditional methods of recording and then interpreting. By adding a crab to the cage she was able to show that predators could equally alter the dominance pattern.

Table 7.1. *Changes in dominance in the fauna of an intertidal mudflat in Washington State, USA, caused by excluding the larvae of tube-building polychaetes by the use of cages for a four-month period*

	Percentage abundance	
Experimental condition	Tube-dweller (*Platynereis bicaniculata*)	Burrower (*Armandia brevis*)
No cage	51.17	7.60
Cage	8.62	33.40

Data from Woodin (1974).

Not all populations show large variations in numbers over time. The burrowing decapod crustacean *Calocaris macandrae*, off the Northumberland coast, has had over a 10-year period a population density of 13.7 m^{-2} with a coefficient of variation of only 5% (the coefficient of variation is defined as S.D./x × 100%). *Calocaris* is territorial and so presumably if one animal dies it is rapidly replaced and the territorial spacing maintained to give the same density per square metre. Similar effects can be expected in other territorial species. Within the community associated with *Calocaris*, however, the other species showed marked fluctuations.

Large areas of the continental shelf of North America are dominated by holothurians, such as *Molpadia oolitica*. *Molpadia* occurs over 440 km^2 in Cape Cod Bay at densities of around 6 m^{-2}. *Molpadia* feeds by ingesting sediment, and thus eliminates most infaunal species; the only common co-occurring species are the tube-building polychaetes and a caprellid amphipod, which by means of its gripping legs attaches to the polychaete tubes. Both the polychaete and amphipod can only exist on the faecal mounds produced by *Molpadia*, since these are relatively stable and are not reingested. Thus *Molpadia* controls its own environment and persists at fairly constant densities from year to year, because it structures its own community. Such species can in fact be called 'key' species, by analogy with the terminology used for rocky-shore habitats (where, for example, the echinoderms such as *Asterias* in Europe and *Pisaster* on the west coast of North America control the community structure by ingesting the potentially dominant competitor for space, the bivalve *Mytilus* (*edulis* or *californianus*, respectively)).

Both of the patterns described above – the oscillating dominance and the persistence through time – are in all probability mediated by biological interactions. In the former case it seems that changes in competitive interactions, geared to the short life-cycle of the organisms concerned, together with seasonal effects such as autumn storms, produce the patterns at frequencies of less than a year. In the latter case the dominant role of the territoriality of *Calocaris* or the sediment-reworking of *Molpadia* produce constant numbers of the individual species concerned by means of competition for space. This pattern can occur not merely within a year but over many years. Other species within the same community, however, may not show the same patterns. Probably the commonest patterns found in benthic communities are the year-to-year ones associated with seasonal changes.

7.2 Annual patterns

Seasonal cycles in the sea influence species even down to the depths of the deep sea. Recent evidence has shown that there are annual breeding cycles at depths of over 5000 m in some ophiuroids, although it must be said that many deep-sea species do not show such patterns. However, annual variations in temperature, light, primary production, etc. can be expected to have a great influence on species occurring in shallow areas, and indeed it seems that the shallower the benthic community the greater the fluctuations it exhibits.

Fig. 7.3 shows data from 50–80 m depth in a community from sand–silt off the coast of Northumberland (Buchanan, Sheader & Kingston, 1978). The pattern is of a clear annual cycle, with an increase in numbers and biomass in the summer and autumn followed by a large winter mortality. The overall mean density and biomass remained roughly constant over the 4-year period. Many similar examples could be shown of such repeatable annual cycles, but these data have been particularly well analysed. The main reason for the large annual changes in the total community is the larval recruitment that occurs in summer. Typically, many species recruit annually, with a large overproduction of larvae that results in a massive mortality produced by competitive interactions or pre-

Fig. 7.3. Annual variation in benthos off Northumberland. (From Buchanan *et al.*, 1978.)

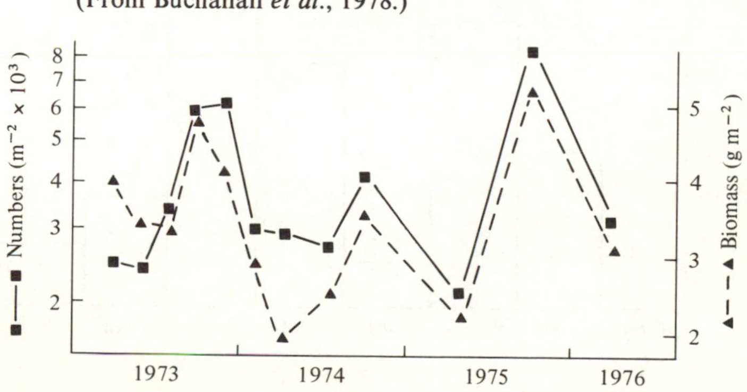

dation together with natural environmental influences. This, then, is probably the commonest pattern that occurs in benthic communities with fairly repeatable annual cycles.

Not all species, however, show annual recruitment patterns; some recruit much more sporadically and this can produce patterns that vary over many years.

7.3 Long-term patterns

One of the advantages of studying population dynamics in bivalve molluscs is that a single sample can, if the species is long-lived, give a picture of the recruitment and growth rates of the species for some years in the past. This is because many species in boreal and arctic waters show growth checks by which annual cycles can be determined and the individuals aged. Indeed, with present-day techniques it is often possible to determine daily growth rates over many past years. Fig. 7.4 shows population figures for the bivalve *Tellina tenuis*, taken from exposed beaches in Scotland. Recruitment was low in 1957, 1961 and 1964, and further data showed that there was no recruitment between 1964 and 1968. In this species, therefore, one can expect that population density will fluctuate quite widely depending on whether or not recruitment occurs. There is a suggestion that recruitment varies with water temperature, but this has not been confirmed.

In the benthic community off the coast of Northumberland the

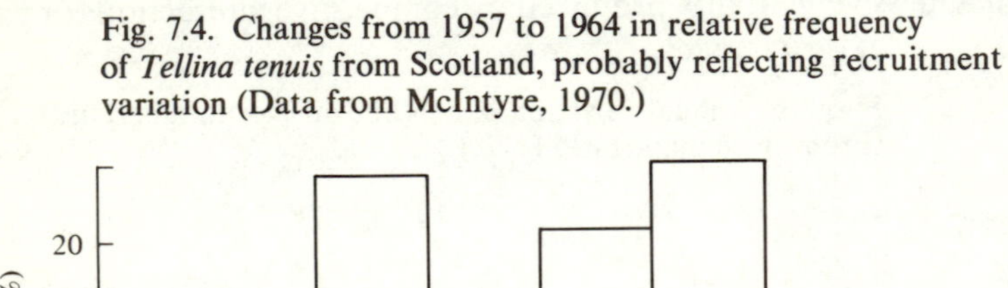

Fig. 7.4. Changes from 1957 to 1964 in relative frequency of *Tellina tenuis* from Scotland, probably reflecting recruitment variation (Data from McIntyre, 1970.)

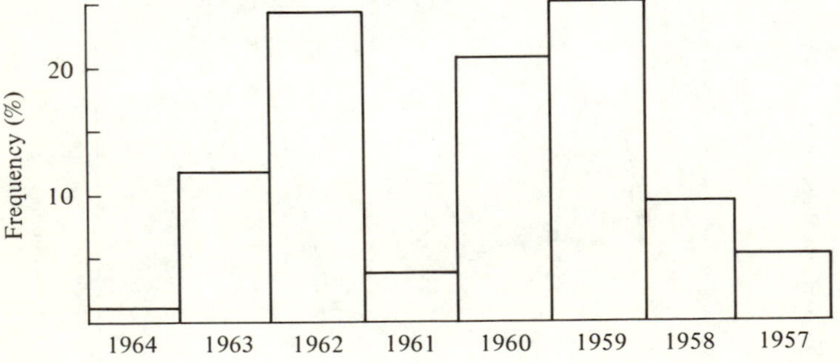

long-term changes were clearly associated with changes in water temperature. Between the winters of 1969–70 and 1970–1 the average winter water temperature rose by over 1 °C. As a consequence, from this point on some species declined in numbers whilst others increased. Fig. 7.5 shows the data. Of the 18 most dominant species, 8 decreased and 7 increased, but the net overall effect on numbers was that the total remained almost the same. Structural changes occurred, but biomass, numbers and production remained constant. It is not known how long this trend will continue, but it is under study.

Another long-term study, but on a single population, is that done by Segerstråle in the Gulf of Bothnia (Baltic Sea) on the relict amphipod *Pontoporeia affinis*. Fig. 7.6 shows the population cycle, which has a periodicity of around 6–7 years. Remarkably, a similar 6–7-year period is beginning to emerge from a study of benthic meiofaunal diversity done off the Belgian coast. It is likely that such

Fig. 7.5. Long-term trends in the benthos off Northumberland. (After Buchanan *et al.*, 1978.)

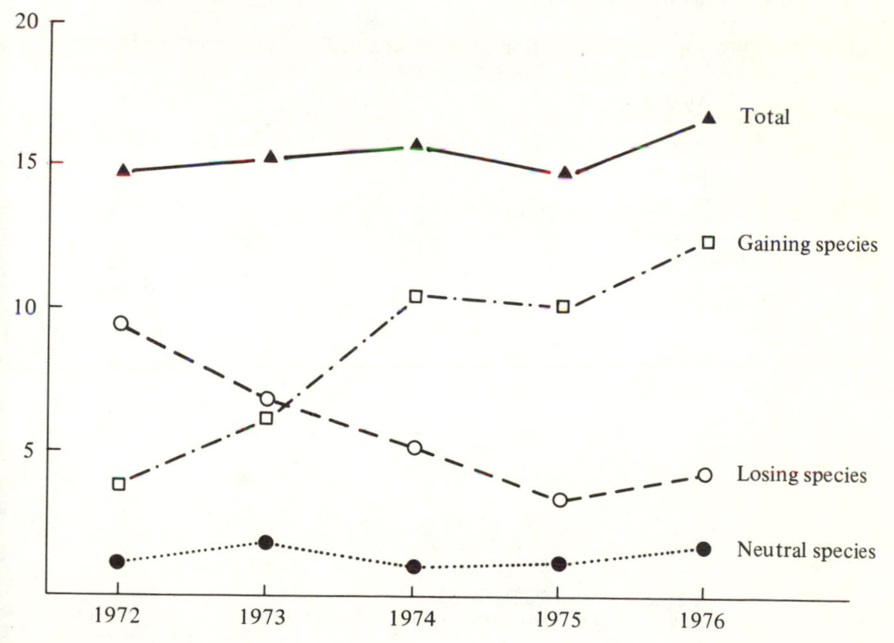

long-term trends will be shown to occur in many species. In the plankton of the English Channel and the North Sea, cycles of 20–30 years are known and seem to be related to long-term changes in the weather patterns, particularly the wind directions associated with changes in pressure fronts. It is too early to say whether or not the same patterns occur in the benthic data, but it seems likely.

The Baltic data on *Pontoporeia* have been recorded since the 1920s and show a pattern of oscillating dominance with the bivalve *Macoma balthica*: when *Pontoporeia* dominates, *Macoma* is scarce and when *Macoma* dominates, *Pontoporeia* is scarce. The period of dominance of each species is around 6–7 years and the change in dominance pattern is mediated by competition for space. *Pontoporeia* adults outcompete the larvae of *Macoma* by destroying them in their feeding activities (not by predation) and *Macoma* can only recruit when *Pontoporeia* densities are low. So here is an oscillating pattern which occurs over long time-scales, not within a year or annually, as shown in earlier examples. Similarly in a pre-pollution study of a benthic community in a Scottish sea loch, numerical dominance oscillated during a 4-year period between four species out of a list of over 140.

Fig. 7.6. Population fluctuations in *Pontoporeia affinis* below 70 m in the Gulf of Bothnia (Baltic). (From Lassig & Lahdes, 1980.)

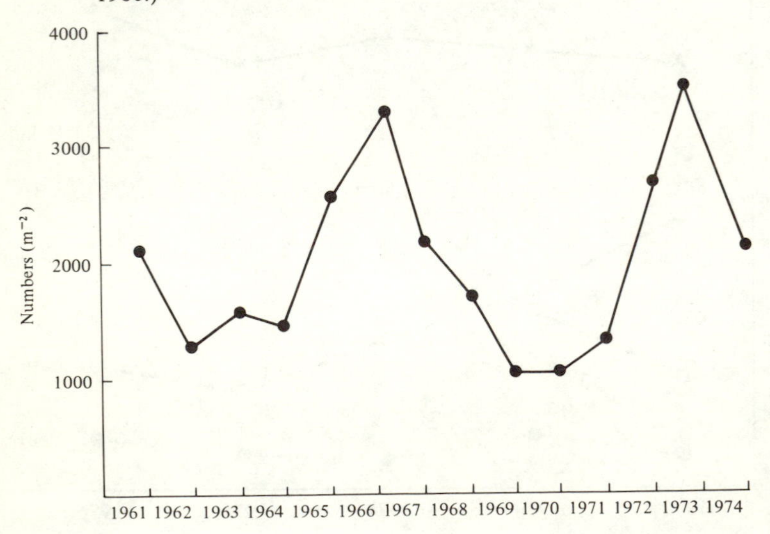

The patterns discussed here have been assessed subjectively. Very recently, however, statistical methods of analysing cycles have been applied to benthic data. Since an increasing effort is being given to the monitoring of long-term cycles, this branch of research is likely to be a fruitful one.

7.3.1 *Analysis of cycles in benthic data*

Cyclical phenomena can be analysed by assuming that the cycles are waves of characteristic length and period. In the previous section the length of the period was the main point of interest. One of the easiest methods of analysing waves to allow estimation of the period is by harmonic analysis using the Fourier series. The simplest of all harmonic series is a sine wave, which has a smooth symmetrical form. Such simple forms are, however, seldom found in nature, and the Fourier series can be used to analyse mixed sine waves of more complex patterns by adding successive components to the original sine wave to give a better fit to the data. The components are simple fractions of the original period and if continued for sufficient terms will fit any data. The object is to add enough terms to give a good fit to the observed points so that the remaining scatter is essentially random. (Bliss (1970) gives a full account of the statistical methods employed.)

Stephenson (1979) in applying the Fourier series to benthic data uses a simplified form of the theorem:

$$y = y^i + C \cos (2\Pi \frac{t - t'}{T}),$$

where y is the population of a given species,

y^i is the time for completion of a complete wavelength,

t' is the lag,

C is half the amplitude of the curve, and

$2\Pi t'/T$ is the time delay in attaining the maximum value.

Before the equation can be applied T needs to be determined. The population at a given time is regressed on its value at a preceding time interval and the correlation coefficient determined. This is called autocorrelation. With a lag of 1 the values of y used will be y_1 and y_2, y_2 and y_3, and so on. The process is repeated for increased lag size and a point will be reached where successive lags are no longer significant. The values of y can then be plotted against the lag values (t') and the peaks observed. For example for a perfect sine curve with wavelength (T) of 12, lag (t') of 9, half amplitude (C) of 10 with a mean of 10, use of the equation above gives:

$$y = 10 + 10\cos(2\Pi\,(t - 9)/12).$$

Fig. 7.7 shows a plot of these data. If a linear trend is added then Fig. 7.7(*b*) results, and finally if a curve of one quarter the wavelength is added ($T = 3$) then Fig. 7.7(*c*) results. One important part of using the above methods is to test how much of the variation is due to a particular cycle. This can be done by using a stepwise multiple regression programme and determining the proportion of variance explained by the regression (r^2). Table 7.2 shows data for the benthic community off Queensland, Australia, sampled 14 times over 2 years by Stephenson.

The trend fitted follows that of Fig. 7.7(*b*), with a sinusoidal wave and a superimposed linear trend. The annual pattern is the sine wave and the remaining long-term linear trend is added. In fact the long-term trend does not have to be linear but may in fact be a wave of extremely long period which over the short time-interval investigated appears linear.

Over all 14 species the proportion of variance due to annual changes was 0.66 for the first year, 0.69 for the second and 0.49 for both combined. There was a long-term linear trend (as in Fig. 7.7*b*) of 0.13, 0.11, and 0.13 of the variance over the same time-periods. Clearly, then, the annual cycles are more consistent within any one year than when combining cycles for two years. One species may have a distinct annual cycle in only one of the two years studied, as did the polynoid (e.g. year 2, 0.82, but year 1 only 0.18). The combined figure for this species is accordingly low (0.16). Some species do not exhibit annual trends at all, for example the sigalionid and the decapod crab *Xenophthalmus pinnotheroides*.

Another aspect that can be obtained from the same analysis is the timing of the maximum amplitude of the curve. In the above data the timing of the cycles within the year varied from species to species, demonstrating perhaps the well-established successional sequences. The timing of the maximum also varied from year to year. Whilst most species showed peaks within a few weeks of one

Fig. 7.7 Theoretical sinusoidal curves generated from a modified Fourier theorem. (*a*) Perfect sine curve; (*b*) sine curve with linear trend; (*c*) sine curve with one-quarter of wavelength added. (After Stephenson, 1978).

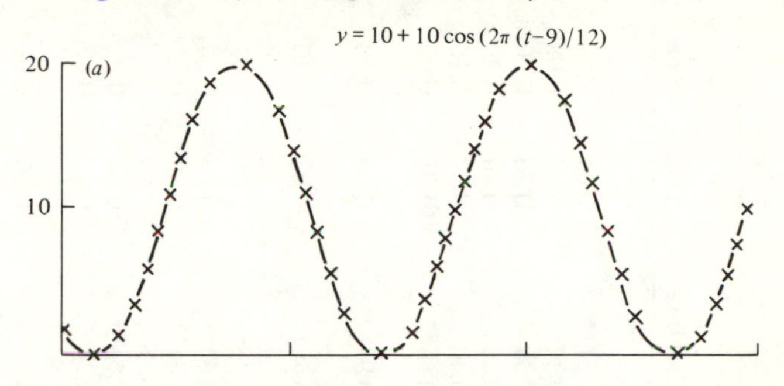

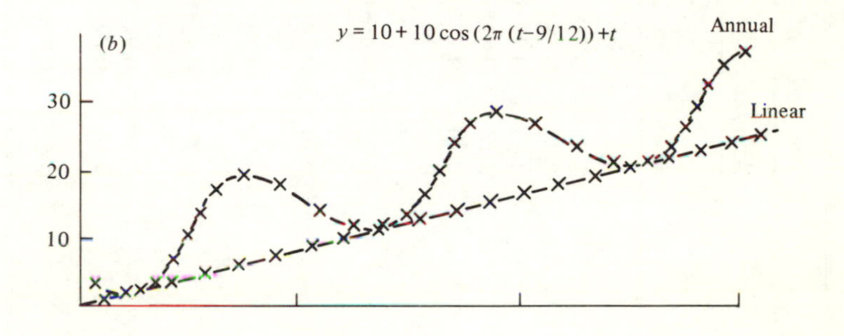

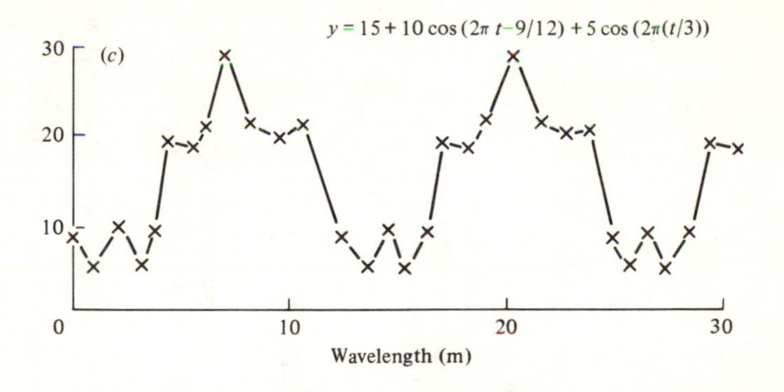

Table 7.2. *Fourier analysis of cyclical patterns in benthos off Queensland, Australia, based on a 2-year survey*

Species	Proportion of variance explained by various components					
	Annual			Linear		
	Year 1	Year 2	Combined	Year 1	Year 2	Com-bined
Polychaeta						
Barantolla lepte (Capitellidae)	0.82*	0.81*	0.61**	0.04	0.00	0.02
Maldanid A	0.75	0.75	0.24	0.04	0.08	0.43
Maldanid B	0.67	0.93**	0.39	0.20	0.00	0.40
Polynoid C (Aphroditidae)	0.18	0.81*	0.16	0.20	0.29	0.04
Sigalionid (Aphroditidae)	0.30	0.27	0.10	0.27	0.16	0.17
Owenia fusiformis (Owenidae)	0.96**	0.74	0.68**	0.01	0.00	0.09
Bivalvia						
Nucula sp.	0.95**	0.91**	0.91**	0.00	0.00	0.03
Scaphopoda						
Dentalium spp.	0.76	0.27	0.46*	0.03	0.20	0.09
Amphipoda						
Amphipod 14	0.78*	0.79*	0.45*	0.00	0.02	0.21
Birubius spp. (Phoxocephalidae)	0.92**	0.31	0.45*	0.03	0.26	0.21
Platyischnopus spp. (Haustoriidae)	0.75	0.93**	0.83**	0.09	0.01	0.00
Decapoda						
Alpheid A	0.48	0.91**	0.66**	0.40	0.02	0.00
Xenophthalmus pinnotheroides (Pinnotheridae)	0.18	0.29	0.21	0.27	0.62	0.10
Foraminifera						
Discobotellina biperforata (Astrorhizidae)	0.65	0.92**	0.70**	0.16	0.01	0.04

Data from Stephenson (1978).
Significance levels: *, $P = 0.05$; ** $P = 0.01$. n.s., not significant.

another, some species had peaks differing by three months from one year to the next. Taking all species together or all stations together there were no significant annual or long-term linear trends. This closely parallels the study off the coast of north-east England by Buchanan, where some species increased and others decreased in abundance, thus giving an overall zero trend (Fig. 7.5). The dominant cycles shown by most species are, however, annual and the timing of these can vary from year to year.

The cycles demonstrated in Stephenson's study are over relatively short periods, since the data were obtained for only 2 years. In a study off the Belgian coast the biomass and diversity of meiofauna have been shown by using similar methods to have 1- , 3- and 6–7-year cycles. The techniques outlined here are, therefore, important tools in quantifying the subjectivity involved in interpreting the cycles discussed in the previous section, and can be expected to lead to a better understanding of long-term variations in benthic communities.

7.4 The stability of benthic communities

Two general patterns for populations emerge from the foregoing. There are some populations, such as *Calocaris* and *Molpadia*, which maintain constant numbers through time, that is they are *persistent*. The other common pattern found in benthic species is that of repeatable *cycles*, either annual, or long-term with periods of 6–7 or 20–30 years. Both these types of pattern can be regarded as being stable since the changes through time are, within certain limits, predictable. The limits vary with the vagaries of recruitment, and may be quite narrow in regularly recruiting species but very wide in species such as *Tellina*. Perhaps, therefore, *Tellina* populations are more unstable, but this really depends on the repeatability of the cycles and there is insufficient information available on this aspect. In fact, so few data are available on long-term cycles and variations in recruitment that the patterns described above may in time prove not to be typical at all. Understanding recruitment variability and the factors causing that variability is one of the central problems in arriving at an understanding of long-term fluctuations in benthic communities.

The other aspect considered was the pattern of variability of species within communities. It was suggested that two general patterns were typical for benthic communities. Firstly, there is persistence, as shown by communities dominated by species such as *Molpadia* that rework the sediment and thereby control the range and numbers of coexisting species. Secondly, there are the oscillating patterns shown by *Nucula*/*Nephthys*, *Ampelisca*/*Nassarius* and *Pontopereia*/*Macoma*. The suggestion made was that these dominance patterns can be changed by environmental factors (*Ampelisca* to *Nassarius*), competition for space (*Nassarius* to *Ampelisca*) or preda-

Fig. 7.8. Models of the stability of communities. The ball represents the community, which can be perturbed from stable equilibrium. (*a*) In neighbourhood stability there are many locally stable points (A–E). (*b*) In global stability there is one unique stable point.

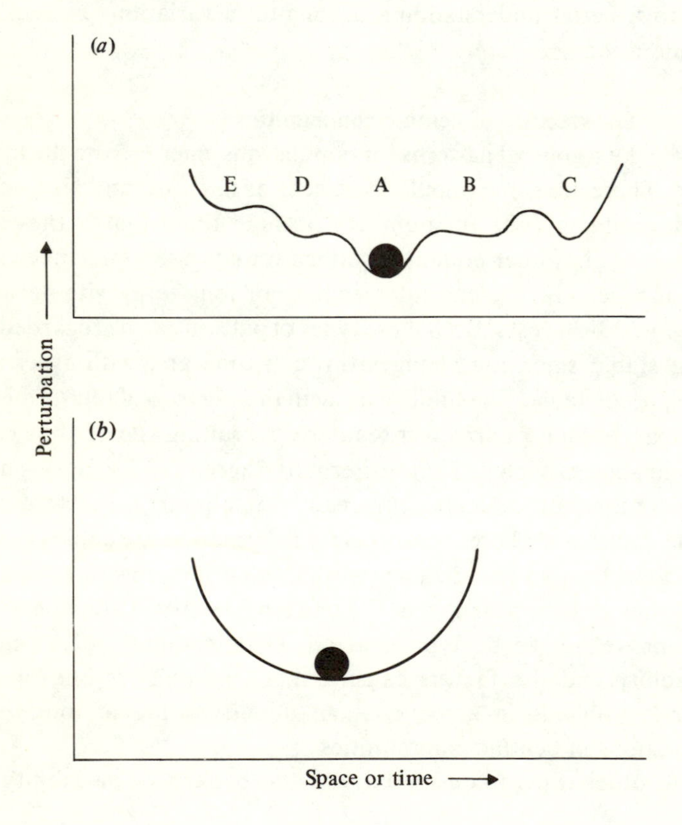

tion (removal of the potential competitive dominant by crabs). Physicists call stability of this kind – where the system is resistant to small changes but moves to another 'basin of attraction' (here another species dominating) on a larger disturbance – neighbourhood stability. This is illustrated in Fig. 7.8(*a*). The community is envisaged as a ball in a basin of attraction, here with species A dominating. The community is resistant to small changes and A continues to dominate. If the environment changes enough, though, the ball may be forced into another basin of attraction where species B dominates, and a predator might shift the dominance pattern to C or back to A. Although the diagram is shown in two dimensions, there may of course be many dimensions, and a species may be highly resistant to, say, a salinity change but easily affected by competition; thus the basin of attraction would be shallow in the competitive axis and deep in the salinity axis. This model, therefore, seems to represent quite adequately many cases of local temporal changes in benthic communities: at one point in time species A dominates, but is replaced by species B which may then go back to A or on to C, depending on which factors are operating. In plankton ecology the same trends have been found, with local temporal patches of dominance occurring, and in this field the glorious name 'contemporaneous disequilibrium' has been applied. This implies that there are local-scale patches which have fluctuating dominance patterns over time. Thus, the neighbourhood stability model applies not only to temporal scales but also to spatial scales. At one point in space species A dominates, whilst a few metres away species B is dominant, and so on. It is a well-known fact in benthic ecology that adjacent grab samples often differ more than do two samples further away from each other, and this reflects contemporaneous disequilibrium occurring over small spatial scales. The patchiness of benthic species in thus part of the expectation from acceptance of neighbourhood stability as a realistic model for some benthic communities (Gray, 1977).

Physicists have an alternative to neighbourhood stability, and that is global stability. In global stability the system always returns to the same equilibrium point no matter how large the disturbance. This is illustrated in Fig. 7.8(*b*). In ecological terms this means that the community always returns to the same equilibrium point with the same species dominating. The difference between neighbourhood and global stability is, therefore, a question of the scale of the disturbance. Is global stability a realistic model for ecological systems? The data in the previous section are not appropriate for testing this model. The ecological analogy here will be that if a storm or a transient pollution incident (oil) grossly disturbs the community, will the system return to the same dominance patterns?

In terrestrial systems the sequence of colonisation has been well studied, and it has been found that one species follows another in a set successional sequence. This sequence was thought by one group of workers to lead to a 'climax community'. Under a given set of environmental conditions one species was thought to prepare the way for another species until finally the succession culminated in the climax where one or occasionally two species dominated. The succession could take many years, but was directional and entirely predictable in outcome; if an oak forest was destroyed by a forest fire then over many decades an oak forest would reappear. However, another group of workers believed that the end-point of the succession was any one of a group of four or five species, and that the sequence leading to these species varied from place to place and time to time. The outcome was not as predictable as was the climax, and was termed the poly-climax.

The colonisation sequence in benthic communities has been studied only rarely. The pattern seems to be that if one ignores the initial microbial aspects, then the first microscopic species to colonise the sediment are the so-called r-selected opportunist species. Such species obtain their name from the logistic equation for population growth:

$$\frac{\mathrm{d}N}{\mathrm{d}t} = rN\left(\frac{K - N}{K}\right),$$

where r is the intrinsic rate of natural increase,
N is population size,
K is the asymptotic density or carrying capacity of the environment, and
t is time.

From this MacArthur & Wilson (1967) suggested that there were basically two extreme types of life cycles. When the population *N* is very small compared with *K*, then *r* is the main determinant of population size and so species with attributes ensuring a high value of *r* will be selected for – hence *r*-selected species. As colonisation proceeds, competitive ability becomes increasingly important as *K* is approached, and so species at later successional stages – so-called *K*-selected species – have as their main attribute competitive ability.

Wherever such colonisation sequences have been followed in benthic communities the initial colonisers always seem to be the same, namely the polychaetes: first *Capitella capitata* (family Capitellidae) and then members of the family Spionidae, often of the genus *Polydora*. Fig. 7.9 shows a typical colonisation sequence. After a relatively short period of time these *r*-selected species are replaced, according to the theory of *r* and *K* selection, by supposedly better competitors. Whether the replacement sequence is in fact a competitive replacement has not been adequately tested and it may well be that the sequence merely reflects the cycle of larval availability and mortality. *Capitella* is able to produce larvae all the year round and can reproduce by both planktonic and benthic phases. Thus, should any space become available, *Capitella* can occupy it. Then, using its other *r*-selected attributes of rapid reproduction and

Fig. 7.9. Recolonisation sequence following an oil spill in Massachusetts (Station B1 at 3 m depth). (Data from Grassle & Grassle (1974); not all species are shown).

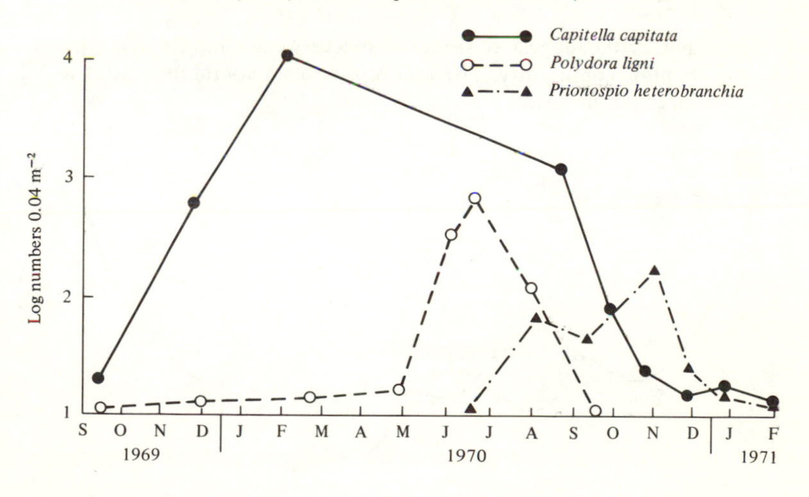

completion of its life cycle within three weeks, it builds up a large population in a short time. Densities of up to 200 000 m^{-2} have been recorded. *Polydora* has many of the same characteristics as *Capitella* (a flexible life-history strategy, short life-cycle, etc.), but is not quite the opportunist that *Capitella* is and therefore comes later in the colonisation sequence. But it is not known why the large *Capitella* population declines. As mentioned above, the classical argument would be that *Capitella* is outcompeted, but no one has shown this. It may just be that the population becomes senile and dies down, leaving space for later colonists that have a more limited breeding cycle with larvae only available over relatively short periods. The initial stages of the successional sequence always seem to follow the pattern above, with the same species dominating; indeed, these species (or near relatives) are cosmopolitan. From here on, though, the sequence varies from time to time and especially from place to place, and no fixed rules apply. Thus, benthic communities follow a successional sequence that parallels the poly-climax idea in terrestrial systems. Indeed, the poly-climax with an unpredictable end-point seems to be a general ecological rule.

The conclusion from these examples, then, is that the early stages of succession in benthic communities are predictable and the same species dominate almost universally, but that later in the sequence the pattern is more varied. This is illustrated in Fig. 7.10. The whole valley is globally stable and the species and relative abundances are

Fig. 7.10. Suggested model of successional changes in a typical benthic community. The ball represents a community after a large perturbation.

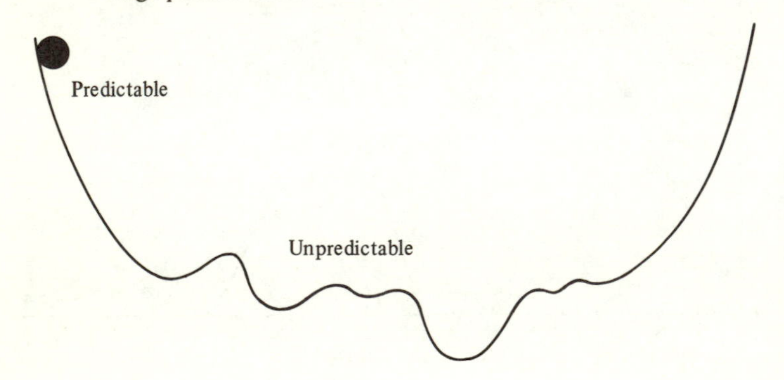

predictable. However, while early stages (the steep slopes of the valley) follow a set pattern, the floor of the valley is bumpy and neighbourhood stability is the rule here, which means one cannot predict the abundance patterns of species at any given point in space or time. This may seem confusing, so let us take a simple analogy from rocky intertidal shores. Given the particular exposure and aspect of the shore an ecologist will be able to predict fairly accurately the abundance patterns and distribution patterns to be expected over the shore as a whole, i.e. the globally stable valley. But the ecologist will not be able to predict which species will occur in a given square metre of shore because on the small scale spatial and temporal variations are in neighbourhood stability (the bumps on the valley floor), and the ecologist cannot know the historical events that have happened at any given point.

Finally, we have to consider how global stability relates to populations. Taking an example already used, we know that if the density of *Calocaris* is reduced from the equilibrium level of 14 m^{-2} to 10–11 m^{-2}, then the density will return to the 14 m^{-2} level since the coefficient of variation is so low over long time-periods. Thus, *Calocaris* responds to small disturbances by returning to equilibrium, that is it shows neighbourhood stability. We do not know what will happen if we reduce the density to 1–2 m^{-2}. Will the population die out or return to 13 m^{-2}? In the former case it would not be globally stable and in the latter it would.

The two concepts of neighbourhood and global stability, therefore, are not mutually exclusive; one or both may apply. Thus the physical concepts are difficult to use in dynamic ecological systems. The acceptance or rebuttal of either model depends on the extent of the disturbance and the scale of the effects.

7.5 Diversity and stability

One of the central ecological debates in recent years has centred on the relationship between diversity and stability. Many years ago the great English ecologist Charles Elton (1966), in his 'girder concept', first suggested that the more diverse a community was the more stable it was. Elton likened a food web to a building with girders. If one or two of the basal girders are removed from a complex food web then the structure will remain almost undisturbed, but if the same is done to a simple food web the structure will collapse; simple systems are, therefore, unstable.

Much effort has recently been concentrated on mathematical models of simple and complex food webs. These studies have shown that there is not necessarily a link between high diversity and high stability (see May (1975) for a review of such aspects). Often the reverse may be true and simple systems can be more stable than complex ones. Although this debate has been going on for some time, the implications for benthic ecology seem to be largely unknown.

A good example of this is provided by a recent debate in Sweden on the funding of marine research. A large-scale and highly detailed study of community dynamics in the Baltic Sea has diverted most of the money available there for marine research to the Baltic. The argument used in support of this was that because the Baltic has low salinity and low diversity hence it is a tough environment for either marine or freshwater species to cope with. Also, because of its low diversity it is unstable and more likely to be affected by pollution than the more diverse west coast of Sweden; a small additional stress factor may be the straw to break the camel's back and cause the Baltic Sea to die. The converse argument was that the brackish-water fauna of the Baltic is physiologically adapted to tolerate fluctuating temperature, salinity and oxygen conditions and is, therefore, probably pre-adapted to tolerate pollution stress. Thus on this argument the low-diversity Baltic Sea is likely to be more tolerant to pollution than the more diverse fauna of the west coast of Sweden. This was the argument used by the west-coast biologists in their attempt to get a larger slice of the research funding! So the debate on the link between diversity and stability can have potentially far-reaching consequences.

This was how I interpreted the two sides of the argument and I thought that the west-coast biologists were correct in that slight pollution would have more marked effects in reducing diversity on the Swedish west coast than it would in the Baltic. But the argument is not quite as simple as that. The Baltic biologists do not dispute the above. They instead ask what happens when the west-coast fauna is reduced to the few species that you find in the Baltic; will an extra stress destroy the fauna of the Skaggerak and Kattegat?

The answer is, clearly, no, since such heavy pollution occurs only in a few limited areas. Yet a similar stress in the Baltic may destroy the complete ecosystem, as is witnessed by the fact that 100 000 km^2 of the Baltic sea-bed below 70 m is anoxic and devoid of life. Here is a telling point in the diversity–stability relationship. Clearly, the open question left by the mathematical modellers as to whether high diversity may or may not be related to high stability has to be studied in each case. Here, therefore, is a highly important research field that has not really been tackled in a comparative way that can provide answers.

In summary, then, marine benthic communities typically show a poly-climax, and neighbourhood stability with a number of alternate dominant species when measured over a small spatial scale or over time. When measured over large spatial scales global stability may be the rule, but since most investigations are over relatively small spatial scales, neighbourhood stability is probably the more appropriate model to be applied.

Populations of territorial species and constant competitive dominants show persistence stability, whereas most marine species typically show cyclical oscillations and 'bounce-back' stability. There is no clear relationship to be expected between diversity and stability, but this has not been studied in a comparative way in benthic communities.

8

Effects of pollution on benthic communities

8.1 Effects on numbers and biomass

Probably the most universal pollutant affecting marine benthic communities is excess organic matter, which occurs principally as sewage but can also include waste from paper pulp-mills, etc. Sewage discharged into confined bodies of water frequently leads to the well-known symptoms of eutrophication, resulting, in the most extreme cases, in a total lack of oxygen and the presence of hydrogen sulphide in the sediment, with a corresponding absence of fauna. As one moves away from the source of pollution there is typically a sudden and rapid increase in biomass and abundance of the fauna. Fig. 8.1 shows data from Kiel Bay. Here an outfall (50 000 $m^3 d^{-1}$) produces an effect only up to 1 km; populations are normal beyond this limit. The restriction of effects to such a relatively

Fig. 8.1. Effect of sewage on numbers and biomass of the benthic fauna of Kiel Bay. (Data from Anger, 1975, in Gray, 1979*a*.)

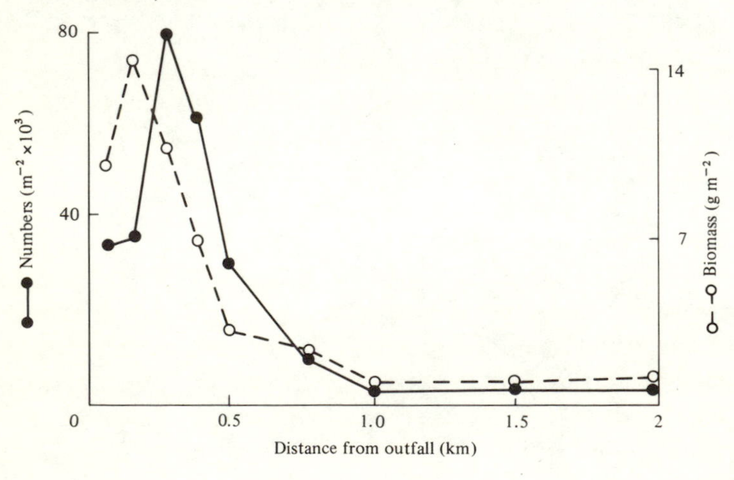

small area is probably due to tidal currents, that both wash away some organic matter and renew the oxygen supply. There are strong tidal currents in many areas of Britain, with the result that effects of outfalls are often confined to areas within only a few hundred metres of the source. By contrast in extremely sheltered areas with little water exchange, such as the Olsofjord, effects of eutrophication stretch over many kilometres. Indeed, Britain has used such facts in opposing the uniform effluent standards that the EEC are trying to impose for the whole of Europe. To the bureaucrats in Brussels it must seem eminently reasonable to insist on uniform standards. But why should Britain put in very expensive treatment plants if the tidal currents give good dispersal and there are no harmful ecological effects? On the other hand, a tourist beach in Italy where there are almost no tidal currents to use must clearly have an efficient treatment plant.

In arguing that it is possible for effects of sewage pollution to be confined to relatively small areas I used the parameters numbers of animals and biomass, but these are rather crude ways of assessing the effects of pollution. One might argue that if more sensitive techniques were used then effects may be found much further away from the outfall. Diversity indices have been seized upon as a sort of panacea in this context. Following the hydroengineering practice of producing simple indices for complex phenomena, administrators and legislators concerned with pollution problems have jumped at the idea of an index which integrates all the species on the interminable lists and tables of numbers that the biologists produce. The naively held view is that here is a simple measure of biological wellbeing; if the index is high there is no pollution and all is well, if the index falls then one should be concerned. Is this view really tenable in practice?

8.2 Effects on diversity

In the chapter on diversity (Chapter 7) I pointed out that there were two aspects of diversity to be considered: species richness (H') and the evenness component (J). Evenness is the inverse of dominance, and so here I shall use $(1 - J)$ and term this dominance.

Fig. 8.2 shows the effects of sewage on diversity (H') and dominance $(1 - J)$ of benthic fauna in Kiel Bay. Diversity is extremely low near the outfall, but rises rapidly so that at a distance of 700 m it is back to its 'normal' level. Dominance is the mirror image of diversity. Thus, *H* is merely reflecting changes in the dominance pattern. A diversity index in this case is no more sensitive than the total abundance or biomass pattern in detecting the effects of pollution, and is considerably more time-consuming to obtain.

One of the most detailed studies yet done on the impact of pollutants on a benthic community is that of Tom Pearson (1975) of the Scottish Marine Biological Association's laboratory. Studying a sea loch, Pearson has followed the changes in the benthos over a 10-year period, for 4 years before discharge of waste began from a pulp-mill and for a further 6 years after. Fig. 8.3 shows the changes in diversity and dominance he found. From 1963 there was a gradual decline in diversity. When pollution began in 1966, diversity continued the steady decline of pre-pollution years. The question is at what stage can we say there is a clear effect of pollution? Certainly by 1969 diversity was very low and pollution was having an effect, but was 1968 part of the pre-pollution trend of falling diversity or not? The interpretation is highly subjective and can be argued both ways. Again, dominance is the mirror image of diversity; when diversity is high dominance is low, and at minimal diversity domi-

Fig. 8.2. Effect of sewage on diversity and dominance of the benthic fauna of Kiel Bay. (Data from Anger, 1975, in Gray, 1979*a*.)

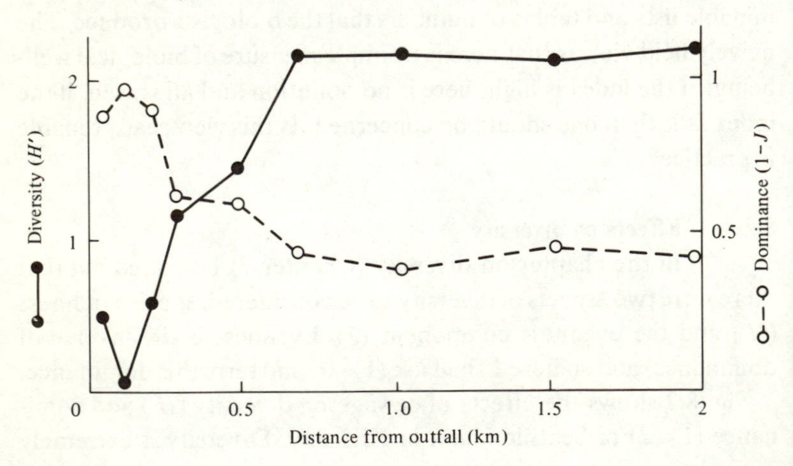

nance nearly reaches the maximum of 1, when only one species would be present.

Interpreting the changes in values of diversity is, therefore, rather difficult. This is in fact not really surprising if we think back to the factors that influence diversity. Diversity is affected by changes in competition between species, by variations in predation pressures, by variations in structural heterogeneity of the habitat and by alterations in environmental predictability; it also alters over evolutionary time. Unless all of these factors remain constant from one sampling period to the next, then one cannot conclude that any change in diversity observed is caused by pollution. Certainly diversity is lowered by severe pollution stress compared with control areas or years, but a diversity index does not appear to be a sensitive tool for measuring pollution effects. In my experience, for a change in diversity to be statistically significant it would have to be equivalent to losing half the species (given initially the distribution of individuals among species and number of species of a typical benthic community). No index is needed to show that half the species are missing and the changes are so obvious that the index tells you nothing!

Pearson also tried plotting Sanders' rarefaction diversity index from his data on the sea loch, but that method does not indicate that significant changes occurred within the first 4 years of pollution either.

Fig. 8.3. Changes in diversity of benthic fauna of a Scottish sea loch on addition of pulp-mill waste in 1966. (Data from Pearson, 1975.)

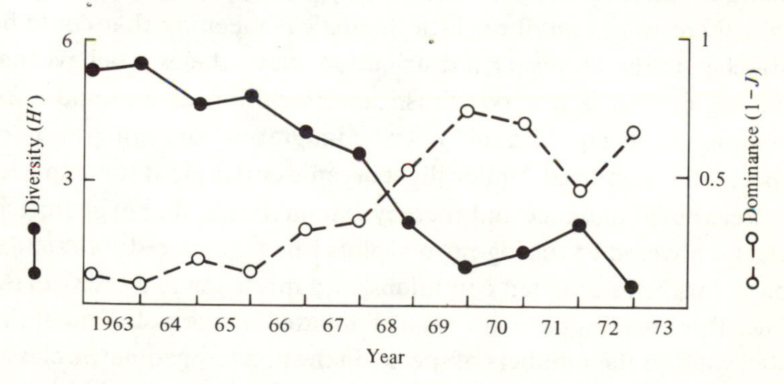

As an alternative approach, I have made log-normal plots of Pearson's data. Fig. 8.4(*a*) shows two of the data-sets of the pre-pollution period (1963 and 1966). There is a perfect fit to the log-normal and the data cover 5–6 geometric classes only. Pollution began in 1966, and already the 1967 and 1968 data (Fig. 8. 4*b*) show a change in the log-normal distribution. The lines have a distinct break and the data stretch over 7–8 geometric classes. Under severe pollution, the line returns to a log-normal distribution by 1970 (Fig. 8.4*c*), but the data now span 14–15 geometric classes and the line has a shallow slope. The above data are from a temporal series; the same pattern also occurs along a spatial series.

Fig. 8.5 shows data from F. B. Mirza, obtained along a gradient of organic pollution in the Oslofjord (Gray & Mirza, 1979). Under the most polluted conditions in the inner part of the Oslofjord, log-normal plots of benthic community data give a shallow line with the data covering many geometric classes. Further away from Oslo a break occurs in the line and the number of geometric classes covered is less. Finally, in the unpolluted areas outside the Drøbak sill, a normal unpolluted pattern of a steep slope with data covering only 5–6 geometric classes occurs. I have plotted similar patterns for many sets of data and am confident that here is a method that is sensitive to changes caused by organic pollution and can be used on almost any community. The log-normal plots detect a change after only 1 year in Pearson's data, which other diversity methods were not able to do with confidence. The important question nevertheless remains: why have these effects occurred?

In Chapter 3 it was suggested that the log-normal distribution could be derived using a variety of hypotheses, and it was argued that there was, therefore, little biological meaning that could be attached to the fit to such a distribution. Nevertheless I believe that the log-normal is a good statistical description of an unpolluted community at equilibrium, where immigration and emigration of species has stabilised. Under slight organic enrichment some species increase in abundance and thereby extend the number of geometric classes covered by the log-normal plots when compared with similar plots for the unpolluted conditions, and this leads to a break in the line. Rare species have not been eliminated by the pollutant at this stage and so the numbers of species in the first few geometric classes

Fig. 8.4. Log-normal plots of benthos from a Scottish sea loch. (*a*) Unpolluted phase; (*b*) transition phase; (*c*) polluted phase. Geometric classes on × 2 scale. (Data from Pearson, 1975.)

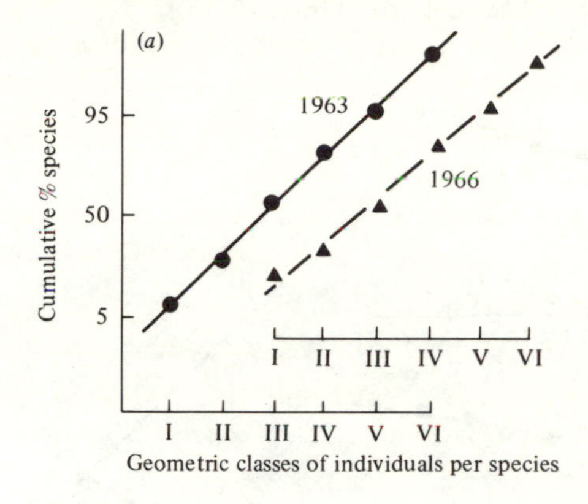

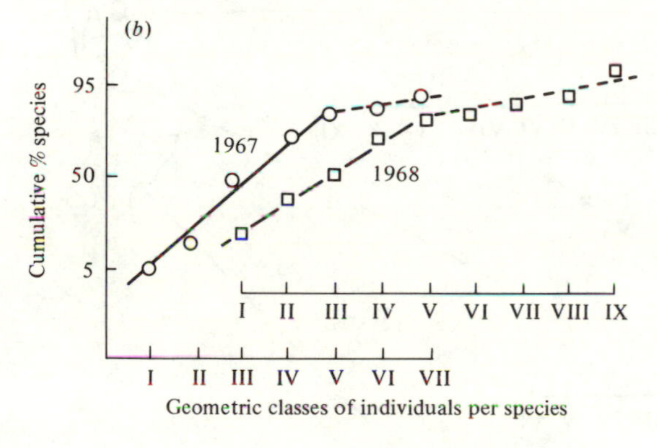

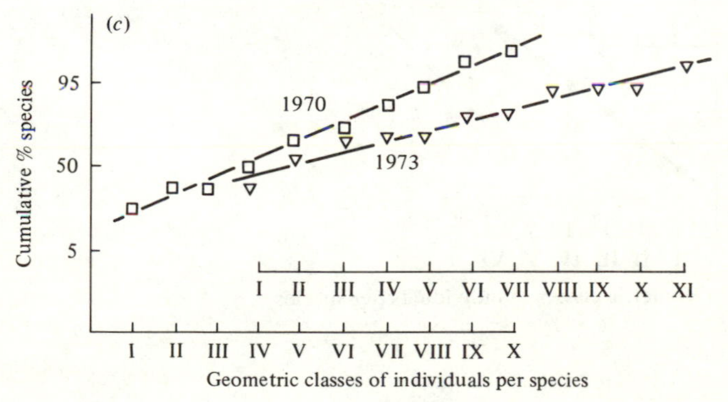

Fig. 8.5. Spatial pollution gradient in the Oslofjord illustrated by log-normal plots of individuals per species against cumulative % species. All data for 0–40 m. A, Bunnenfjord, polluted; B, Vestfjord, transition stage; C, unpolluted; D, unpolluted geometric classes on × 2 scale. (After Gray, 1979*b*.)

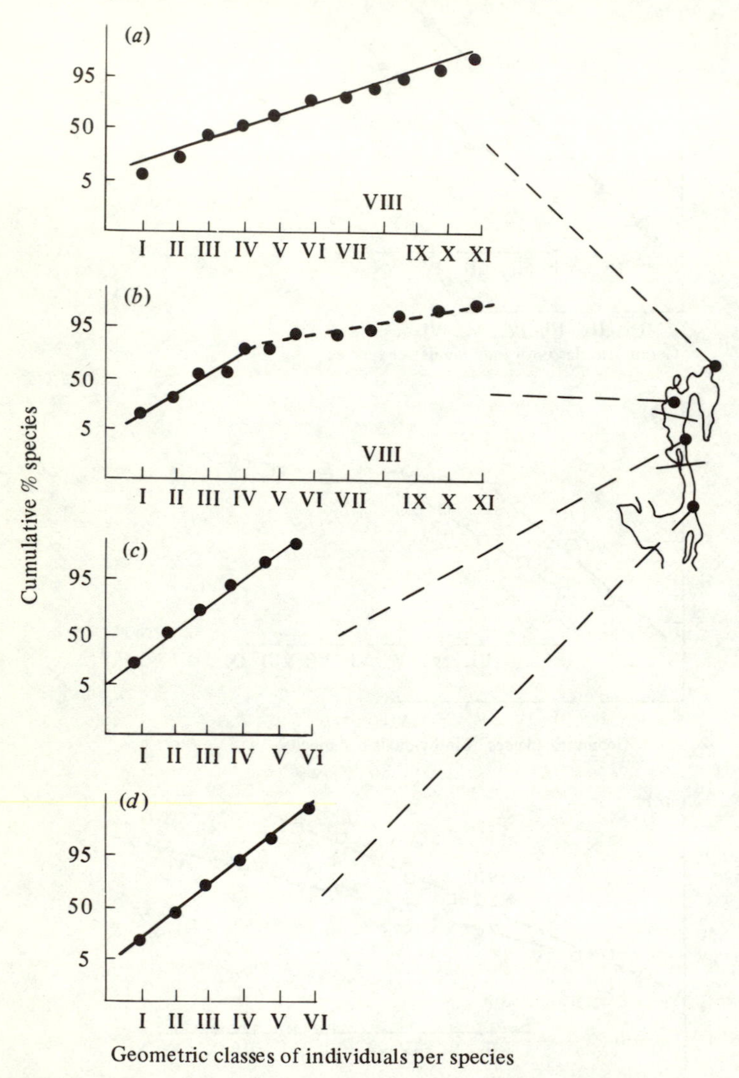

remain the same retaining the original slope of the line. As pollution increases an equilibrium community is re-established, with some species becoming extremely dominant and most of the rare species being eliminated, which gives a shallow slope to a log-normal plot that now covers many geometric classes.

The first detectable change is suggested as being a break in the straight line of the plot, which represents a departure from an equilibrium community. There is no reason to suppose that such a change need be brought about by pollution. Indeed it has been shown that a storm or even seasonal recruitment of larvae produce a break in the log-normal line similar to that in Fig. 8.4(*b*). However, unlike communities affected by pollution, communities affected by a storm or larval recruitment will return to the unpolluted type of log-normal line after a relatively short time. Thus a break in the log-normal line is merely suggestive of a non-equilibrium condition; only if it is long-lasting is it likely to indicate an effect of pollution. (This point will be taken up later when I consider monitoring methods.)

Theoretical arguments supporting my hypothesis that the log-normal represents an equilibrium community have recently been provided by Nils Stenseth (1979), based on Van Valen's Red Queen hypothesis (see p. 28). Stenseth provides many examples showing the good fit of the log-normal distribution to species abundance patterns from many varied communities. He goes one step further by proving mathematically – and supporting with empirical data – that in disturbed communities a log-series distribution is the better fit.

8.3 Species changes under pollution

So far I have not mentioned species changes caused by pollution. Pearson & Rosenberg (1978) have just completed a thoroughly comprehensive review of the effects of organic pollution on benthic communities. Using data from many case studies they have produced a diagram representing pollution effects on a boreal fjordic community (Fig. 8.6). There is a sequence of changes. Firstly the most sensitive species are lost; then as the RPD layer approaches the surface and pollution becomes more severe the community is progressively simplified to only a few species.

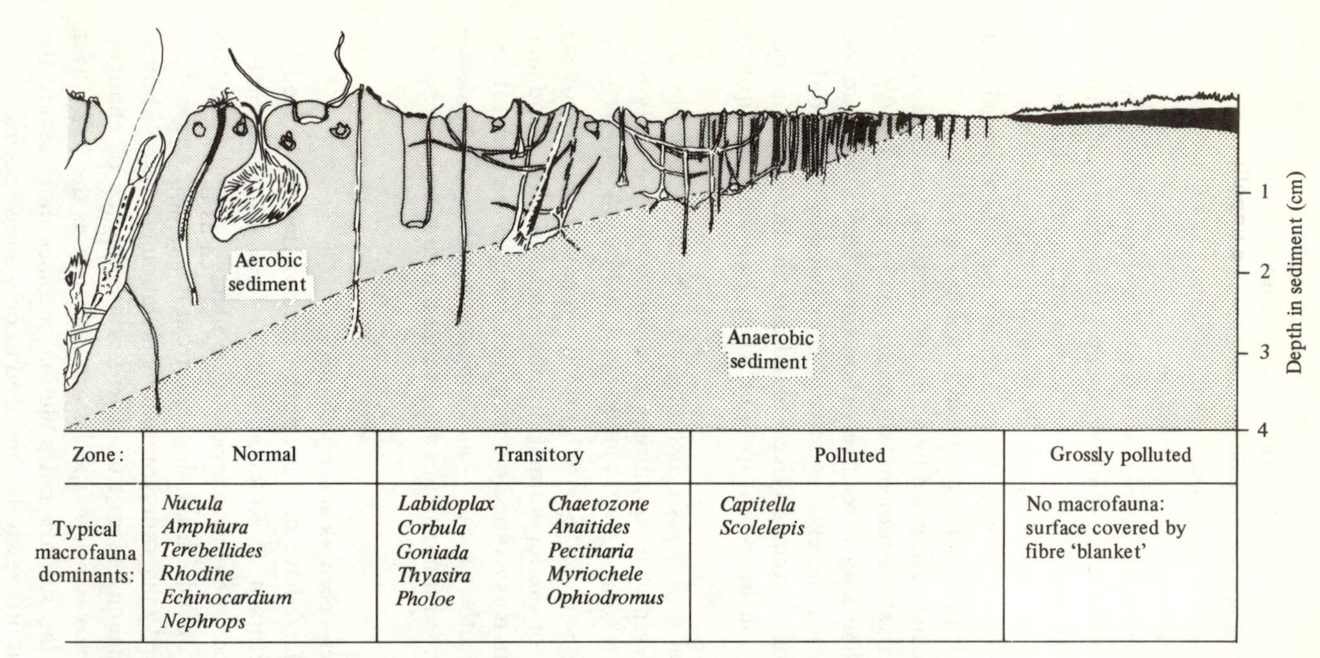

Zone:	Normal	Transitory	Polluted	Grossly polluted
Typical macrofauna dominants:	*Nucula* *Amphiura* *Terebellides* *Rhodine* *Echinocardium* *Nephrops*	*Labidoplax* *Corbula* *Goniada* *Thyasira* *Pholoe* *Chaetozone* *Anaitides* *Pectinaria* *Myriochele* *Ophiodromus*	*Capitella* *Scolelepis*	No macrofauna: surface covered by fibre 'blanket'

Fig. 8.6. Diagram of changes in fauna and sediment structure along a gradient of organic enrichment. (After Pearson & Rosenberg, 1978.)

In a typical benthic community there are something in the order of 150 species of macrofauna. Fig. 8.6 gives some important species from this large total which can be used to indicate the changes occurring. It seems to me more realistic to use such groups of species to monitor changes in community structure than to try to find a single 'indicator species'.

A similar short-list of species can be obtained from log-normal plots. The break in the log-normal line was suggested as being due to a number of species which increased in abundance. If my interpretation is correct then those species are the same ones as those categorised as transitory by Pearson & Rosenberg. The main difference between the two ideas is that the log-normal method is a general method applicable to whatever community is studied, whereas the list of transitory species is for the boreal region only. It is rather like the difference between the Petersen–Thorson idea of communities being rigid descriptions and the concept of a continuum: the modern idea gives more flexibility. As the log-normal plots apply to any community, it is not expected that the species lists produced

Table 8.1. *Species in geometric classes V–IX which increase in abundance under slight organic pollution*

Species	Unpolluted	Polluted
Oslofjord		
Cirratulus cirratus	0	83
Glycera alba	64	107
Pectinaria koreni	0	103
Polyphysia crassa	0	95
Sabella pavonina	2	57
Thyasira spp.	12	687
Scotland		
Cirratulus cirratus	7	30
Pholoe minuta	17	35
Prionospio cirrifera	40	290
Corbula gibba	20	35
Lucinoma borealis	11	38
Thyasira flexuosa	96	441
Labidoplax buski	193	226

Data from Gray & Mirza (1979) and Pearson (1975), respectively.

from different plots will be the same, even within small areas. Table 8.1 shows the species occurring in geometric classes V–IX (i.e. those classes where the break occurs) of the log-normal plots from both Pearson's data from Scotland and Mirza's data from the Oslofjord. The close similarity with the species in the transitory zone shown in Fig. 8.6 is evident. Yet the list derived from the log-normal plot did not require any special ecological expertise, whereas Fig. 8.6 is based on many years' experience by practising ecologists and a thorough literature review.

8.4 Adaptive strategies to pollution

Producing a list of the different species that occur along a gradient, whether of salinity in an estuary or of organic pollution, is a typical so-called ecological approach. Few people have attempted to ask the important question of why the pattern occurs. Why is it that species A can survive whereas species B disappears? Why, for example, does the polychaete *Capitella capitata* occur in the most organically polluted areas? The standard answer is that *Capitella* is the most tolerant species – more tolerant of lower oxygen tension and higher hydrogen sulphide than any other species – and that such conditions are to be found in the most polluted areas. In fact, *Capitella* is not particularly tolerant of low oxygen tension, and in comparative studies three other polychaetes typical of polluted areas were found to be more tolerant. Yet it is *Capitella* that is consistently found in the most polluted areas. So an explanation other than tolerance must be found.

I believe that pollution effects can loosely be separated into two categories: disturbance and stress (Gray, 1979*a*). By disturbance I mean an effect whereby individuals are physically destroyed or removed from an area. Chemical stress, on the other hand, results in the productivity of an individual being reduced. The adaptive strategies shown by organisms to these two factors are illustrated in Table 8.2.

Organisms adapt to high chemical stress by increasing their tolerance. Under high disturbance the best adaptive strategy is to have a high *r* value (see p. 82 for *r*-selection), by having a rapid reproductive rate and turn-over time, reaching maturity rapidly, and being relatively short-lived. Under normal conditions without pollution stress *r*-selected species are gradually outcompeted and replaced by *K*-selected species, which reproduce more slowly, are slow to reach maturity and are long-lived. There is no adaptive strategy to a combination of high stress and high disturbance and under such conditions no animals can exist.

Sewage pollution is primarily fine, particulate matter falling onto the sediment. The particles are rich in organic matter and bacterial activity is consequently high. If the sewage loading is high, many species are simply smothered by the falling particles and cannot survive. *Capitella* survives because it is a classical *r*-selected species: it can reproduce both by planktonic larvae and by benthic larvae, has a short life-cycle, and reaches maturity from the egg in about three weeks. It can, therefore, continuously repopulate sediments subjected to pollution from organic matter. *Capitella* does not use tolerance as an adaptive strategy, but adapts to continuous disturbance by continuous reproduction.

An oil pollution incident occurred in Massachusetts, USA, in 1969. In the recovery sequence *Capitella* was the first species to recolonise and rapidly built up a large population, reaching densities of over 200 000 m^{-2}. Then just as rapidly the population declined and was replaced by another polychaete, *Polydora ligni*. The oil killed the natural fauna and in the disturbed environment *Capitella* had the ideal life-history strategy to recolonise, by means of the ever-present planktonic larvae. Once the population was established, it could be built up rapidly by reverting to benthic larval production. Yet, according to the theory of *r*- and *K*-selection, *Capitella* is not a good competitor and is therefore ousted by *Polydora*. However, I must add that there is no direct evidence that *Polydora* does in fact outcompete *Capitella*. It may well be that the *Capitella* population becomes senile and *Polydora* occupies the

Table 8.2. *Adaptive strategies to pollution*

	Low stress	High stress
Low disturbance	Competitive (*K*)	Tolerant (*T*)
High disturbance	Reproductive (*r*)	Non viable

From Gray (1979*a*).

newly available space. The competitive exclusion theory is, however, more satisfying and it should be a simple matter to verify. Interestingly, *Polydora* also shows many attributes of an *r*-selected species, but it is clearly not quite so opportunistic as *Capitella*. *Polydora* usually reproduces by means of a planktonic larva, but can also brood within its tube, again a life-history strategy ideally adapted to disturbance. *Capitella* has been found in high numbers following a tropical storm, a red tide, sediment excavation, etc., and indeed is to be expected wherever the sediment is newly disturbed.

In contrast to species adapted to disturbance, stress-tolerant species are slow-growing and of low competitive ability. Arctic and desert plants are typical examples of tolerant communities, but others are possible (see p. 62, and Grime, 1979). If the environment becomes less stressful, then these *T* species will be outcompeted by *K* strategists, which are also slow-growing but are better competitors.

Capitella has been suggested as a 'universal indicator of organic pollution' (i.e. where it is abundant there is likely to be organic pollution). Yet as can be seen from the above examples *Capitella* is abundant only in disturbed areas, such as those affected by a storm or an oil spill. And the large *Capitella* populations that build up following these disturbances are transient; as colonisation proceeds *Capitella* is outcompeted and returns to its normal low population densities. It may indicate organic enrichment, however, if it continuously occurs in large numbers, but even then it is only 'may' indicate. For example, some recent data from a dock in Merseyside, England, showed that in an area subjected to regular sand spillage (i.e. disturbance) the numbers of *Capitella* remained high. But it is in fact usually much easier to see that there is a large amount of organic matter in an area than to see large populations of *Capitella*! *Capitella* represents an end-point in the effects of organic pollution, as was demonstrated in Fig. 8.6. What the biologist needs are species that indicate the first stages of decline of an ecosystem. But, nevertheless, there are important clues to be gleaned from studying *Capitella* and trying to discover why it is so successful in coping with pollution, and this I feel is the sorely neglected area in pollution research. So let us return to the transitory species, or those isolated from break-point in the log-normal distribution, and try to find out whether or not they adapt to pollution in the same way that *Capitella* and *Polydora* appear to do.

The examples of pollution-tolerant species that I have used have inevitably biased the data towards pollution in terms of disturbance, since this is the principal effect on communities of both oil spills and organic pollution. Chemical pollution can lead to quite different species abundance patterns from those produced by disturbance, if the effect of the chemical is merely to reduce the productivity of the competitive dominant *K*-selected species without killing it; for while the *K*-selected species physically occupy space the *r*-selected species will not be able to establish themselves. But if the chemical(s) kill species then a disturbance effect will occur and the *r*-selected species can be expected to become established. Thus it is rather difficult to predict the effects of chemical pollution. Remarkably, I have had great difficulty finding good data on which I can test the above ideas. There are almost no published data on the species that are found around chemical plants or on their densities. Presumably most of the relevant information lies in confidential company files! So here again is an important area of research that can be expected to give some interesting data on effects of pollution. My prediction is that it will be extremely difficult to fit the effects of chemicals into general patterns. Some species will be affected by one chemical and not another, and only when species are killed will the *r*-selected species have a chance to establish themselves. Thus, since each effluent varies in chemical composition and concentration, all manner of species combinations can be expected.

Fig. 8.7 integrates the three adaptive strategies shown by species to pollution. At the lower left are those species that, according to the hypotheses suggested, are expected to occur in the most polluted environments. Whilst I believe that these hypotheses will be found to be correct, much more evidence is needed in order to test them adequately. In particular, the life-history strategies and tolerance capacities of the potentially sensitive species need to be studied. There are almost no data on the biology of the species listed in Table 8.1 and this is a research area that a number of students at the University of Olso are following.

Another interesting point regarding adaptive strategies to pollution is the important finding of Fred and Judith Grassle (1973) of the Wood's Hole Oceanographic Institute. Working on the *Capitella capitata* population that followed the Massachusetts oil spill they found that, on the basis of enzyme polymorphisms, rather than being one species there were in fact six sibling species. Using classical cross-breeding techniques they showed that crosses between the species were infertile and that thus they were indeed true species; but they were hard to separate using conventional anatomical characters. The different species varied in type of larval development and length of larval life and in their season of maximal abundance. In fact, the successional sequence that followed the oil spill should probably include a succession of the different species of *Capitella*. In a highly opportunistic species such as *Capitella* selective forces probably lead to a continual evolution and extinction of species. Interestingly, *Polydora*, which also lives in highly polluted areas and is highly opportunistic, is also taxonomically difficult. In the Oslofjord *P. ciliata* shows an interesting temporal adaptation. Populations in the most polluted innermost part of the fjord produce larvae for 12 months of the year, populations 10 km down the

Fig. 8.7. Adaptive strategies to pollution (based on Fig. 8.6).

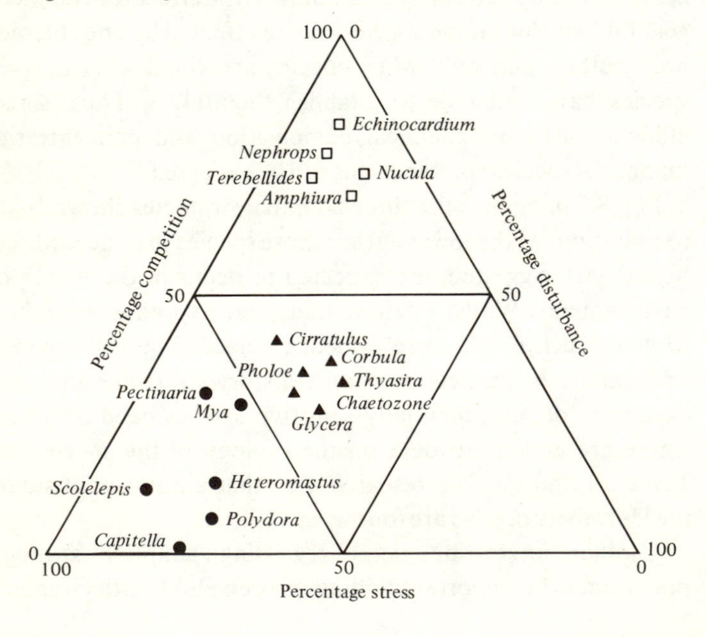

fjord, where it is less polluted, produce larvae for 6 months of the year, and those in the relatively unpolluted area near the inner sill produce larvae for only three months of the year. The life-history patterns of the different species of *Polydora* along the pollution gradient in the Oslofjord are under study, together with the genetic aspects (i.e. degree of heterozygosity of populations) as shown by enzyme polymorphisms, to see if the same pattern found by the Grassles occurs in the Oslofjord.

Two other species typical of polluted areas are, like *Capitella* and *Polydora*, cosmopolitan, and I believe that if they are studied in depth they too will be found to have the same life-history and genetic adaptations as *Capitella* and *Polydora*. The two species are the polychaete *Heteromastus filiformis* and the oligochaete *Peloscolex benendeni*.

We have seen from the above that in relation to organic pollution, at least, the adaptive strategies adopted by species do not appear to be based on tolerance. Yet the classical way of trying to predict what effect chemicals are likely to have in the marine environment is to do toxicity tests. In such tests organisms are placed in various concentrations that are lethal to 50% of the population, usually determined in a 48-hour test. From work done in fresh water it has been found that one-tenth of the LC_{50} value gives a 'safe limit'. Similar standards have been adopted for the marine environment. From the arguments above, it will be seen that the toxicity test does not in fact allow one to predict the ecological consequences of pollution, since the species adapt to increased stress by altering their life-history strategies rather than increasing their tolerance. Thus I question the value of toxicity tests that aim to predict safe limits for discharges to the sea.

Whilst this account by no means exhausts the subject of the effects of pollution on benthic communities, I hope that the ideas presented here will perhaps stimulate new approaches to the topic. I believe that it is important not merely to provide lists of species occurring along a pollution gradient, but also to ask why these particular species are present. What special attributes do these species have that enable them survive? More studies of a natural-history type are needed if we are to answer these questions; it is unfortunate that natural history is unfashionable.

Before leaving the topic of pollution one problem remains to be considered: the long-term monitoring of benthic communities that is frequently done to provide a base-line for the assessment of the effects of pollution. This is the subject of the following chapter.

9

Long-term monitoring of benthic communities

The foregoing chapter was concerned with the acute effects of pollution, when pollution came from a specific source and was easy to see. The more difficult aspect of pollution research is that of assessing the chronic effects of small amounts of pollutant discharged over long periods of time. Here the problem of detection is much more difficult, since one has to separate the effects of pollution from the natural variability in populations. When discussing stability in Chapter 7 I was mainly concerned with relatively short-term changes – oscillating patterns and seasonal cycles – but what do we know of the really long-term variability in benthic populations? The answer, surprisingly, is very little. There has been a great deal of research on benthic communities, but most on a rather sporadic basis with investigations lasting only a few years. Really long-term and truly quantitative data are hard to find.

9.1 How long is long-term?

Two sets of data give us some idea of likely trends. The Baltic studies on *Macoma balthica* and *Pontoporeia affinis* have already been mentioned in Chapter 7. This study has covered many decades and a recent paper suggests that *Pontoporeia* undergoes, in addition to annual cycles, long-term cycles of 6–7 years that are probably related to long-term climatic changes. If such cycles are the rule then the detection of effects of chronic pollution must be followed over similarly long periods or else there is a risk that a decline in some community parameter recorded over 3–4-year period may just be part of the natural cycle. Thus, the first prerequisite of a monitoring programme designed to test the effects of chronic pollution must be that the programme runs over decades. This point is not, of course, appreciated by funding agencies! Until this point is understood, however, one runs the risk of throwing away enormous sums of money on studies that have no value because they have been done over unrealistically short periods of time.

Another unfortunate aspect of so-called pollution monitoring programmes is that they often take no account whatsoever of any theoretical aspects of benthic ecology. A typical approach that illustrates such naivety is to select a given area, take the largest number of samples possible, measure a number of environmental factors, identify and count the macrofauna, and repeat this at monthly intervals for a couple of years. Why do I regard this as a naive approach?

To me the programme above involves an unrealistically and unnecessarily large amount of work. What one really needs to do is to rationalise the programme according to sound ecological arguments.

9.2 Deciding on which species to monitor

If one begins with the species list of a typical benthic community, there are of the order of 150 species of macrofauna and probably twice as many meiofauna. Most of the species are rare, represented by only one or two individuals, as shown by the fit to the log-normal distribution. The problem is that one seldom knows why these species are rare. To me, the rareness of most species is one of the most intriguing problems in ecology. In benthic samples many species are never found at densities greater than 1 or 2 per grab sample, wherever or whenever the samples are taken. Are these species found in high densities elsewhere; or do they always have that spatial pattern because that is their optimal density; or do they occur in these patterns as a result merely of random effects such as larval settlement, predation, or differential environment-caused mortality? Whilst the collective disappearance of a large number of rare species would, of course, be indicative of change, our lack of knowledge of rare species suggests to me that they can reasonably be excluded from a rationalised sampling programme.

But how can we further reduce the list to a few useful indicator species? One possible method has been illustrated (p. 97) that uses the log-normal distribution; it was substantiated by the wealth of data from Pearson & Rosenberg's review of the effect of organic pollution on benthic communities. The problem is, of course, that in an unpolluted area where the interest is in acute effects there will be no break-point in the log-normal line, and it is on this break-point that the method depends. Perhaps all that can be done is to select those species that occur in geometric classes V–IX (i.e. between 16 and 255 individuals per unit area), since the numbers are suitable for the application of statistics and yet the tedium of counting large population densities is avoided. For example, in the polluted part of the Oslofjord densities of *Polydora* approach many thousands per square metre, and following the oil spill in Massachusetts densities of *Capitella* reached 200000 m^{-2}. But since these species are opportunists, they can be expected to show very large variations in population density naturally over relatively short time-intervals. Thus in this context such species are unsuitable, since we are primarily interested in long-term changes and it is a waste of time to count large numbers of individuals that merely show short-term changes.

Another method of direct relevance is to try and ascertain which species control the dynamics of a given community. On rocky shores on the west coast of America, for example, the dynamics of the shore are controlled by the feeding of the starfish *Pisaster* (Paine, 1966). *Pisaster* eats the potentially dominant competitor for space, *Mytilus californianus*, and therefore allows other species such as the gastropods and sea-anemones to colonise the area. This was established by the simple method of erecting cages over the rock and excluding *Pisaster*; the effect was that *M. californianus* rapidly covered 100% of the available surface. Thus *Pisaster* has been called a 'key' species in controlling the dynamics of many rocky-shore species. A similar thing occurs on European shores, but to a less extent, since in Europe the common starfish, *Asterias rubens*, comes onto the shores only in summer and thus has a seasonal predation effect, whereas *Pisaster* is present the whole year round.

Two types of interaction that create 'key' species in benthic communities have already been mentioned: competition (between *Ampelisca* and *Nassarius*), and feeding behaviour (of the holothurian *Molpadia oolitica*). It is remarkable how important the echinoderms are in controlling community structure. For example, on subtidal hard substrata the kelp beds along the Canadian coast, which once covered hundreds of kilometres, have disappeared. It is thought that this is due to overfishing of the lobster, which has allowed enormous populations of the echinoid *Strongylocentrotus droebachiensis* to grow up. The *Strongylocentrotus* has decimated the kelp beds by feeding on them. On subtidal soft sediments there are often incredibly high densities of ophiuroids and undoubtedly these, together with the holothurians, are important in structuring subtidal communities. Again, the interrelationships have been relatively poorly studied.

In communities where such 'key' species can be found a monitoring programme can reasonably be concentrated on them, since they are likely to determine the dynamics of the other members of the community. How widespread communities structured by 'key' species are, remains to be seen from more detailed manipulation experiments. The indications are that, in fact, benthic sedimentary communities do not have the same structuring factors as rocky shores; many more species are present. It is suggested that this large number of species is due to the third spatial dimension afforded by the sediment, which allows many more potential niches and hence species. Thus, monitoring in sediments is likely to involve many more species than on rocky shores. For example, in Europe one could probably construct a valid monitoring programme for a rocky shore by basing it on *Balanus balanoides*, *Patella vulgata*, *Mytilus edulis* and *Asterias rubens*, whereas in a sedimentary habitat a minimum of 9–10 species is likely to be necessary.

As mentioned above, 'key' species can be discovered by manipulating the communities (as will be shown in Chapter 10). There is not the time to wait for nature, and deduce the effects of competition and predation *a posteriori* from many years' data. Manipulation techniques reduce the time-scale necessary to a few months in many cases. Also, the traditional *a posteriori* methods of interpreting complex data-sets cannot give answers, but only generate hypotheses which must then be tested. For example, one might after collecting data for a number of years find that the decline in an abundant species coincides with the increase in abundance of a much rarer predator. One would suggest that this was a predator–prey relationship, but proof would await a thorough test.

9.3 How often should one monitor?

Having established that there are a group of 9 or 10 species that can be followed (for example those in Table 8.1 would be ideal for following the effects of organic pollution along a gradient in the Oslofjord), then the next question concerns the frequency of sampling. Earlier I suggested it was naive to collect monthly samples. Clearly, if there are no data for a given area then one would need to establish the sorts of seasonal patterns that occur and monthly samples might be appropriate. But it may be more valid to take samples at times when the water temperature rises by 1 °C, for example. Or, if one is interested in life history, then it would be better to take few samples in the winter, when, in the case of most species, nothing is happening, and instead concentrate the samples over the spring and early summer when reproduction and recruitment occur.

The main point that I am trying to make is that sampling must always be based on biological criteria, and there must always be a reason for taking a sample. For example, in long-term monitoring are we really interested in the larval settlement period? The time of reproduction of a given species can vary by three to four weeks from year to year, depending on many different environmental factors, and the reproductive output can vary by an order of magnitude. But frequently by a couple of months after the initial settlement large mortality has occurred and the population has returned to a 'normal' level. Fig. 9.1 shows a log-normal plot from samples taken at bimonthly intervals off the German coast over a number of years. Clearly, in summer larval settlement causes the community to depart from equilibrium. The timing of this departure varies from year to year, yet by September all communities have returned to equilibrium. Is it worth counting in a monitoring programme, therefore, all the many hundreds of settled larvae which will die within a couple of months? Obviously not, since we are interested in detecting long-term chronic effects of pollution. Thus the strategy that I would recommend, and one which is being used in the Oslofjord, is once per year sampling in winter when population densities are low and there are no settled larvae to count (which in any case are usually on the border between macrofauna and meiofauna). Once per year sampling is already being done off the coast of Northumberland by one of the most experienced benthic ecologists in

Europe, Jack Buchanan (Buchanan *et al.*, 1978), and I am convinced that his approach of concentrating on the dynamics of 10 to 12 species is the correct one.

To summarise, the monitoring strategy that I would adopt for benthic communities is as follows:

1. Thoroughly describe the species present in a given community and estimate their relative abundances; characterise the habitat by grain size, sorting coefficients, organic content, etc.
2. Rationalise the species list to around 10 important species which control the dynamics of the community. This can be done by use of manipulation experiments where appropriate.
3. Monitor these species by intensive sampling once a year when population densities are low and there is little

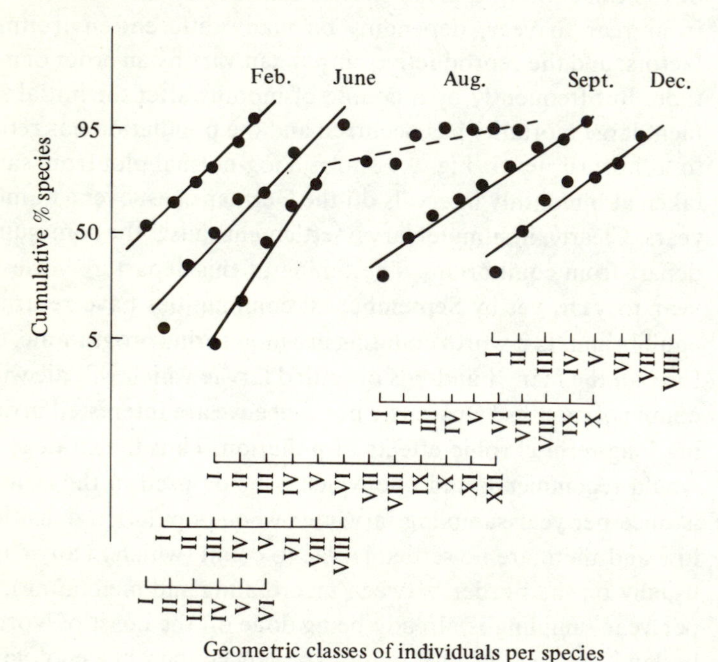

Fig. 9.1. Seasonal changes in log-normal plots from benthic data of Helgoland Bight. (Data from Rachor & Gerlach, 1978, in Gray, 1979*b*.)

> chance of recording larval settlement (in boreal areas this will probably be in mid-winter). Record as many population parameters as is possible for the species studied, such as numbers, biomass, age structure, size frequencies, growth rates, etc., plus as many environmental parameters as is practicable.

It might be mentioned again here that bivalve molluscs are particularly useful in a monitoring context, since they are long-lived and usually have annual growth rings from which it is often possible to infer growth rates for a number of past years which were not sampled. Monitoring by recording growth-rate parameters of a number of species is being done in Shetland by Alan Jones of Dundee University, and looks to be an extremely useful and sensitive method.

I have not dealt with techniques for the monitoring of individuals, since I believe that unless an effect is measurable on populations it has no real ecological relevance. But many new and interesting biochemical and physiological techniques are being developed and tested (see review articles in McIntyre & Pearce, 1980), and the latter are likely to provide an early-warning system that can alert ecologists to the necessity of measuring the effects of some environmental disturbances or stress at the population and community level.

10

Factors controlling community structure

10.1 The trophic-group amensalism hypothesis

One of the most obvious relationships between benthic communities and environmental factors is that between feeding type and the silt–clay fraction of the sediment. Sanders (1958) was the first to quantify this relationship, from his studies of the benthic fauna of Long Island Sound, New York. Fig. 10.1 shows plots of the percentage (in terms of numerical composition) of suspension- and deposit-feeders against the percentage of silt–clay. Suspension-

Fig. 10.1. Relationships between the percentages of suspension-feeders (circles) and deposit-feeders (triangles) and the percentage of silt–clay in the sediment. (Data from Sanders, 1956.)

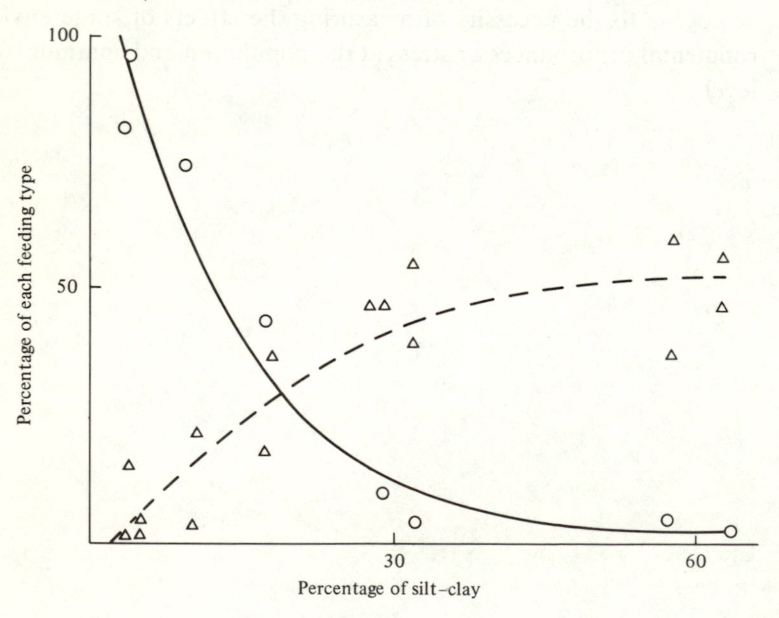

feeders have highest densities in sandy sediments. Sediments with particle sizes of around 0.18 mm are the most stable since such particles are the easiest to transport (p. 11) and, therefore, where they occur wave and current action must be minimal. Sanders predicted that suspension-feeders should exploit this property and have highest densities at this particle size. His data from Buzzard's Bay seem to support this idea. Deposit-feeders, on the other hand, reach maximum densities in muddy sediments, because the organic matter content is high.

The boundary between suspension-feeding and deposit-feeding communities is often sharp; it is also found that hard substrate (e.g. anchors or bottles) placed in muddy sediments rapidly become covered by dense suspension-feeding populations. From this evidence it seemed probable to two American workers, Donald Rhoads and David Young (1970), that deposit-feeders actively inhibited suspension-feeders and somehow prevented them from occupying muddy sediments. Using an ingeniously designed camera that could take pictures of the sediment–water interface and down to a few centimetres into the sediment, they showed that in silt–clay sediments the deposit-feeding bivalves *Nucula proxima*, *Yoldia limatula* and *Macoma tenta* constantly reworked the sediment, turning the silt–clay into faecal pellets the size of sand particles. The depth of reworking could be measured simply by measuring the water content of the sediment, which was much higher in the heavily reworked areas. Fig. 10.2 illustrates a typical result. Based on the reworking data Rhoads & Young produced the 'trophic-group amensalism hypothesis' to explain the absence of suspension-feeders on silt–clay deposits. (Amensalism means that one population (in this case suspension-feeders) is inhibited whereas another (here deposit-feeders) is not.) They suggested that the unstable silt–clay cannot be colonised by larvae of suspension-feeders, and that even if it could their gill structures would be rapidly clogged by resuspended material. More sandy substrate, however, favour suspension-feeders. Here the sediment is firm, enabling their larvae to form attachments to it, and is not suitable for deposit-feeders because there is not sufficient food within the sediment to support them and the coarse particles make locomotion difficult. Furthermore, by an amensalistic interaction suspension-feeders could destroy the larvae of potentially recolonising deposit-feeders. In conclusion, Rhoads & Young suggested that three types of community should exist:

1. homogeneous suspension-feeding groups;
2. homogeneous deposit-feeding groups;
3. mixed communities where limited reworking of the sediment means that the bottom is sufficiently stable to allow suspension-feeders to colonise.

Rhoads & Young were careful to point out that the hypothesis only applies to communities of the continental shelf where primary productivity is high and food is not limiting to suspension-feeders. However, in a follow-up study to the one on which the trophic-group amensalism hypothesis was developed, Young & Rhoads (1971) found that suspension-feeders increased along a gradient of increasing silt–clay in Cape Cod – in direct contradiction to the predictions of the hypothesis. Here the tubes of a polychaete, *Euchone incolor*, stabilised the sediment and allowed suspension-feeders – *Thyasira gouldii* (bivalve) and *Aeginia longicornis* (amphipod) – to colonise in large numbers. The two suspension-feeders capitalised

Fig. 10.2. Effect of reworking by bivalves on the percentage water content of sediments. Values are means of four readings. (After Rhoads, 1974.)

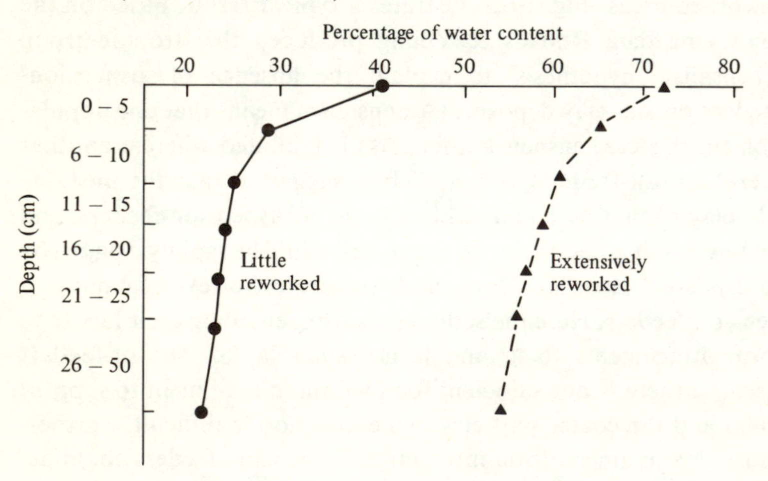

on the stabilised sediment and were thus not subject to the amensalistic effects of sediment reworking. Sediment stabilisation, particularly by polychaetes, is common, and is clearly important in structuring communities and providing exceptions to the trophic-group amensalism hypothesis.

10.2 Food limitation in suspension- and deposit-feeders

The next development in the theories of control of community structure came from Levinton (1972) and extended the above ideas. He suggested that suspension-feeders rely largely on sedimenting plankton for food, for the gut contents of many species reflect what is available in the plankton. Yet plankton is extremely variable, in both space and time, in terms of quantity and the species present. Thus, as suspension-feeders remain fixed to the substrate and cannot move to better feeding areas, they must be typical opportunists, rapidly exploiting favourable conditions by building up large populations – and having equally dramatic population crashes. (It is these species that are commercially exploited.) Levinton argues that it is therefore unlikely that two such species compete for long enough to reach exclusion or to produce niche specialisations, and predicts that suspension-feeders will thus have broad overlapping niches with no feeding specialisations.

Deposit-feeders, on the other hand, probably utilise bacteria as their main food source. And during feeding they produce large amounts of faecal pellets. Fig. 10.3(*a*) shows data on pellet formation in the intertidal mud snail, *Hydrobia minuta*. *H. minuta* will not reingest its own faecal pellets until they have been broken down (Fig. 10.3(*b*) shows breakdown rates of faecal pellets), and thus rates of pellet formation and breakdown can be very important population-regulating factors for the species, since if the sediment becomes almost all pellets then the animals stop feeding. Faecal pellets are extremely common in muddy sediments both inter- and subtidally; In the inner Oslofjord between 70 and 90% of the 'sediment' is pellets. Levinton argues from such data and the known behaviour of *H.minuta* that, whilst there is potentially a large amount of organic matter present in muddy sediments, most of it is not available because it is in pellets that deposit-feeders will not reingest. Breakdown rates of pellets depend on bacterial action, and indications from the nitrogen and ATP contents of muds are that bacterial activity is fairly constant. In such sediments, therefore, the food resource is predictable and limited. Levinton suggests that deposit-feeders have competed for this limiting resource over evolutionary time and that as a result feeding specialisations have evolved. Again, the hypothesis is qualified as only referring to the trophic dimension of the niche and Levinton suggests that the spatial dimension may, in fact, be very important. He also argues that in muddy sediments stabilised by polychaetes, continuous resuspension of material may provide a predictable food source for the suspension-feeders, implying that food specialisations may also occur in such situations.

10.3 Adult–larval interactions

It was the presence in the sediment of a third trophic grouping – tube-builders which stabilise the sediment – in addition to suspension- and deposit-feeders that led Sally Woodin (1976) to propose what she has called an 'alternative' to the trophic-group amensalism hypothesis. She suggests that 'assemblage types result

Fig. 10.3. (*a*) Faecal pellet production by *Hydrobia minuta*. (*b*) Breakdown of *H. minuta* faecal pellets in a laboratory experiment. (From Levinton & Lopez, 1977.)

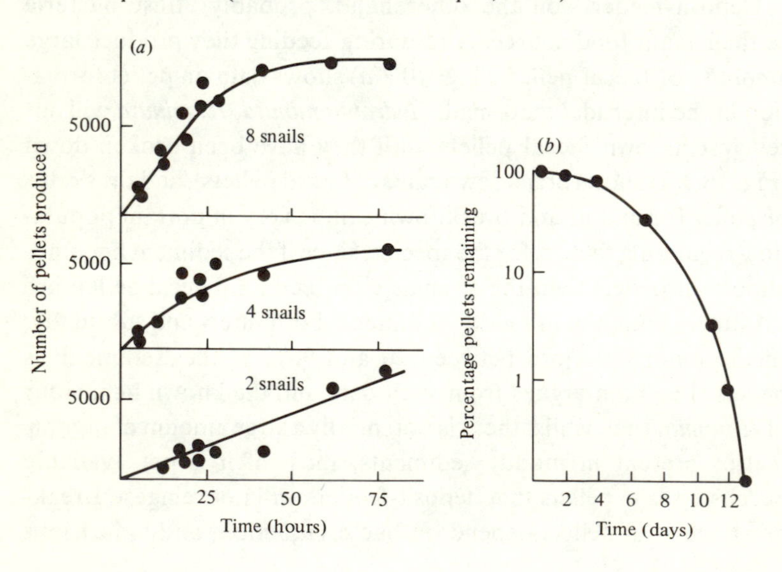

from the interaction between established individuals and newly settled larvae rather than between two established individuals'. Yet this seems to me to be precisely the same argument as that used by Rhoads & Young and indeed discussed by Levinton. In describing trophic-group amensalism Rhoads and Young state that 'The physical instability of the reworked enivronment ... may discourage the settling of larvae of suspension-feeders. If settling does occur, early growth stages may be inhibited or killed by the unstable sediment conditions. The inhibitors (deposit-feeders) are unaffected by this relationship, while the amensals (suspension-feeders and sessile epifauna) are either discouraged from settling or are killed during early benthonic stages.' Thus, I do not believe that Woodin's ideas are a new hypothesis; but she has extended the prediction of the trophic-group amensalism hypothesis. She predicts, on the basis of adult–larval interactions, that:

1. Suspension-feeding bivalves which brood their young and release eggs too large to be eaten by tube-builders should be abundant among tube-builders.
2. Small burrowing polychaetes should reach their highest densities among deposit-feeders, since they are eaten by suspension-feeders as larvae.
3. No infaunal forms should consistently attain high densities among suspension-feeding bivalves.
4. As bivalves are perennials and can thus destroy their own larvae when filtering the water or reworking the sediment, this will give rise to age-class phenomena.
5. Tube-builders, as a result of their dense populations, should also destroy their own larvae when feeding and thus cause similar age-class phenomena, but since they are annuals their population should not persist.

Much theorising about benthic communities has been done within the last few years. I believe that such ideas are necessary to enable us to obtain an understanding of how these communities are structured and function. For too long most so-called benthic ecology consisted of *a posteriori* correlations of numbers with, for example, environmental factors, a method which cannot of course definitely establish cause-and-effect relationships. The hypotheses described above give us insight into the workings of communities, and Sanders, Rhoads & Young, Levinton and Woodin have done invaluable service in stimulating more goal-directed research. But questions remain as to how their hypotheses stand up to counter-arguments.

10.4 A critical look at controlling factors

From the foregoing it appears that there are three very distinct trophic groups in benthic communities, namely (1) deposit-feeders which rework the sediment, (2) suspension-feeders which do not build tubes, and (3) suspension-feeders which build tubes and stabilise the sediment. If such sharply distinguishable communities were common, then a Petersen–Thorson-type classification scheme should work, since consistent patterns and few overlaps would occur. But species are distributed log-normally in continua along environmental gradients rather than being in such clearly defined groups, and thus the trophic groups referred to in the above hypotheses are rather rare and are not the general rule. Sally Woodin referred to this point in saying that on the west coast of North America such groupings were common, whereas on the east coast they were rare, and suggested that the reason for this may be that in east coast communities predators have an overriding effect. Certainly in Europe sharp trophic-group boundaries are the exception rather than rule. *Polydora*, a tube-builder, does build reefs on mud, but not commonly. There are also local patches of dominant suspension-feeders, such as the dense population of *Spisula elliptica* described by Hagmeier in 1925 on the Dogger Bank. But I believe that Rhoads & Young's 'mixed communities' are far more common, and here the dominance pattern may oscillate between a suspension-feeder and a deposit-feeder in the pattern shown in Chapter 7.

There is little doubt that interference competition, where one species prevents another from exploiting a resource in sediments by means of reworking, is a common and potent force in benthic communities. Such interference *may* result in the total exclusion of suspension-feeders from a sharply defined area, but I suspect that it is more likely that it results in their absence only from small-scale patches and causes local temporal changes in dominance patterns (see stability argument in section 7.4).

Levinton's argument, that suspension-feeders are unspecialised opportunists with broad overlapping niches, is convincing. Their food is unpredictable and may occur in super-abundance, so to be an opportunist is the appropriate life-history strategy. There are, however, suspension-feeders which show opportunist-type blanket settlement of larvae but that, unlike classical *r*-selected species, are long-lived. In sediments the cockle, *Cerastoderma edule*, and on rock the mussel, *Mytilus edulis*, fit this category. Both these species are able to adapt to low food levels merely by growing slowly. On the exposed coasts of north-east England, individuals of *M. edulis* measuring only 10 mm in length could be up to 15 years old. Thus, after blanket settlement, high densities are maintained by intraspecific competition for food that leads to adjustments in growth rate. Here, then, is a slightly different 'strategy': instead of a classical *r*-selected opportunist that is extremely fecund and short-lived, here is a species that is physiologically highly flexible in terms of growth rate and fecundity. But such a strategy is in keeping with that of an opportunist, where stress leads to reduced growth; the same occurs in plants (Grime, 1979). And again like an opportunist *Mytilus* still maximises gamete production at the expense of growth (Bayne, 1980); this probably also occurs in *Cerastoderma*. Thus the only real divergence from the classical *r* strategy is length of life.

Food is not the only limiting factor. One can expect and indeed find that there is intense spatial competition in sediments, since niches overlap. No species is ever in the ascendency for long enough to produce niche specialisation or competitive exclusion, but over short-time periods competition can be intense (see stability-time hypothesis, p. 59).

The view that deposit-feeders are food-limited is more equivocal. Levinton's argument that the rate of faecal pellet breakdown in hydrobiid gastropods limits population size is a convincing one. But he argues from this that as hydrobiids produce faecal pellets which are not immediately reingested, then all species must do the same, and I do not see that this should be the case. In fact many deposit-feeders are known to be coprophagic, and often such species play an important role in breaking up faecal pellets into finer particles where bacterial action can be more rapid.

In the Oslofjord faecal pellets do accumulate, and since there are no tides and currents there are no natural disturbing agents such as there are intertidally. But I do not believe that there is any evidence to show that in areas of the fjord where there is high pel-letisation food is limited. In fact, macrofaunal abundances are highest where most pellets are found. Thus Levinton's hypothesis may only apply to rather limited areas.

Levinton is most careful to point out that he is talking only about the *trophic* dimension of the niche, but this has important ramifica-tions. Many of the specialisations shown by deposit-feeders are in the form of segregation at different depths within the sediment, i.e. there is apparently competition for space. But a given deposit-feeder presumably has a certain area that it must use in order to obtain enough food, and this area will be intensively competed for. (A parallel can be drawn with birds: here too territories are defen-ded, the territory also being the feeding area of the individual (or pair); moreover, different species of bird occupy different layers within any given tree species.) Thus, the apparent competition for space is really competition for food, the depth specialisations re-flecting competitive interactions over evolutionary time for food resources. In fact space alone is seldom likely to be a limiting re-source in a three-dimensional habitat. Is food then? I have suggested above that the pelletisation argument is probably only valid in cer-tain areas; the same is likely to be true for food also. My argument depends on data from manipulation experiments, and I will discuss this further in section 10.5.

The first two of Woodin's five predictions – that (1) epifaunal bivalves have eggs too large to be eaten by tube-building suspension-feeders and should reach their highest densities among tube-builders, and (2) small burrowing polychaetes cannot exist among suspension-feeders because they are eaten as larvae and, therefore, have highest densities among deposit-feeders – are both obvious, and examples of each are known. The third prediction, that no in-faunal species should occur in high densities among suspension-feeding bivalves, is contradicted by the abundance of infauna among clumps of *Mytilus edulis*, both intertidally and subtidally. Nereid polychaetes are particularly common among such clumps. The explanation is that suspension-feeders do not cover 100% of the water available (as is well known from studies on filtration efficiencies of bivalves) and so recruitment is clearly always possible for infaunal species. Once within the clumps, there is an abundance of food in the form of pseudofaeces and true faeces, in addition to some sedimenting organic matter.

The last two predictions, of age-class phenomena resulting from the adults present destroying settling larvae of their own species, imply that all adults die simultaneously, thereby giving a chance for complete recruitment of an age-class. But mortality is usually patchy and varies over time, and thus larvae can recolonise and replace dead adults over the whole life-span of the organism. In fact the age-class phenomenon cannot be explained by adult–larval interactions, and is more likely to be the result of good and bad recruitment years in turn linked with climatic variations, such as was suggested in the 'match–mismatch hypothesis' put forward by Cushing (1975) to explain variations from year to year in plankton abundance. The year-class phenomenon could occur if the adults were to render the habitat unsuitable for themselves and thus cause mass mortalities. This has been known to happen in the case of *Spisula elliptica* on the Dogger Bank in 1925, which suffered mass mortalities when its feeding rendered the sediment coarser. The important point here is that it is the organism itself that changes the environment and results in year-class phenomena, not adult–larval interactions.

It is not my purpose here to try to refute all the hypotheses. Indeed, the trophic-group amensalism hypothesis is a convincing one. I am merely attempting to point out that the hypotheses above are not based on all the available evidence and, therefore, seem untenable as universal rules. Clearly, no single hypothesis can explain all the complexities of distribution patterns in benthic communities. What needs to be done is to erect hypotheses that can be tested, and the next section shows that this has been done to good effect.

10.5 Experimental manipulation of benthic communities

The problem with many of the hypotheses discussed in the previous section is that they are extremely difficult to test adequately. How do we test whether food is limiting for deposit-feeders? Does the presence of many small polychaetes among deposit-feeders really substantiate Woodin's adult–larval hypothesis? A great deal has been learnt about how benthic communities are organised by means of experiments that manipulate natural populations *in situ*. The first experiment of this type was done as long ago as 1928 by the Dane, Blegvad, who was interested in what effect fish predators had on the benthos. He put cages over the sediment to exclude the fish, and found that the fish's preferred food was 60 times denser under the cages than in the areas outside. The experiment was done over a summer.

Over 40 years later this experimental technique was revived by Sally Woodin (1974). She was a student of Robert Paine at the University of Washington, Seattle, who had done an elegant and simple experiment of a similar type on rocky shores. By means of cages placed on the shore Paine (1966) established that the community structure was determined by the asteroid *Pisaster*, which ate the potentially dominant competitor for space, *Mytilus californianus*. With *Pisaster* present, many other species were able to utilise the space created by *Pisaster* eating *Mytilus*. In the absence of *Pisaster*, *Mytilus* monopolised all available space. Woodin did similar experiments, using cages to exclude predators, on a muddy intertidal shore. Table 10.1 shows the results of one such experiment. Only

Table 10.1. *Abundances of three tube-building and one burrowing species within 0.05 m² of caged and uncaged sediments*

	Tube-building species				Burrowing species
	Lumbrinereis inflata	*Axiothella rubrocinctata*	*Platynereis bicaniculata*	Total	*Armandia brevis*
No cage	168	123	358	649	47
	92	158	313	563	52
Cage	168	136	47	351	143
	132	153	25	310	160
	113	141	19	273	129
	64	104	54	222	139

Data from Woodin (1974).
Samples taken from August to November–December.

P. bicaniculata and *A. brevis* reproduced over the experimental period. There is a clear change in the dominance pattern, with the burrowing species increasing in abundance within the cage. Woodin suggests that the tube-builders settle on the cage surface while the burrower passes through the cage onto the sediment. Normally, therefore, the tube-builder outcompetes the burrower for space by means of interference competition. In her experiments Woodin demonstrated two other important things: that when a predator (a crab, *Cancer magister*) is placed inside a cage the abundance of tube-builders decreases dramatically whereas the burrower maintains the same numbers; and that variations in larval recruitment can strongly influence the community structure found. These findings stimulated a whole host of similar experiments.

In attempting to improve on Woodin's experimental methods I decided to use a large cage within which I could subsample at regular time-intervals. Whereas Woodin's cages were only 28 cm × 34 cm, my design was 2 m × 2 m. The cage was placed on an intertidal area of the Tees Estuary, England, where there are large populations of overwintering bird predators. No significant changes in species abundances occurred, probably because the large size of the cages meant that it was impossible to exclude, for example, buried crab predators.

One of the most dramatic series of predator exclusion experiments that has been done is that of Reise (1977), working on the intertidal sandflats of the German Wadden Sea. He put down cages for varying time-intervals and found the results shown in Table 10.2. Not only did the total number of individuals rise by factors of up to 20-fold, but also the number of species increased. Indeed, such effects seem to be common trends in all caging experiments done at shallow depths: numbers, biomass and number of species all increase under the cage. The last point is in marked contrast to similar experiments done on rocky shores, where typically competitive exclusion is the rule, the number of species under the cage falling as one species, usually a mytilid, outcompetes other species for space. This is one of the crucial differences between the sediment and rocky-shore experiments. But why should competitive exclusion occur on rock and not in the sediment?

The argument that when predators are excluded on rocky shores

Table 10.2. *Abundances and number of species of macrofauna within 400 cm^2 of caged and uncaged areas in an intertidal sandflat, Wadden Sea, Germany, after three months*

	June to Oct. 1974		Mar. to June 1975		July to Oct. 1975	
Species	Uncaged	Caged	Uncaged	Caged	Uncaged	Caged
Capitella capitata	7	32	41	222	37	56
Polydora spp.	0	71	15	104	10	213
Peloscolex benendeni	65	1792	180	373	328	1222
Heteromastus filiformis	52	45	47	25	96	89
Pygospio elegans	1	157	237	706	7	140
Tharyx marioni	9	889	63	203	3	2129
Cerastoderma edule	2	189	20	307	3	513
Macoma balthica	1	5	28	939	0	2
Total no. individuals	150	3459	828	3332	477	4937
Total no. species	12	22	21	25	7	28

From Reise (1977).

the number of species decreases, is based on cages which exclude *Pisaster*; *Mytilus* then dominates the newly available space. The success of *Mytilus* is primarily due to it having larvae and juveniles available to colonise the space and to its ability to overgrow other potential space-occupiers. The crucial question that needs to be asked is whether the species number *really* declines. In the sediment experiments all species retained on a 1 mm or 0.5 mm screen are counted, but the rocky-shore data of Paine and others refer to the occupancy of space by species that can be observed directly. There is a rich infauna within *Mytilus* clumps which is not taken into account in the estimates of species numbers. Here a number of new, rare species could be expected to occur within the increased area of the clumps, just on the basis of species/area relationships. The rocky-shore data, therefore, specifically exclude the third spatial dimension of depth, and refer only to competitive exclusion in two dimensions; it is not possible to do this in a sedimentary environment. Thus, the data are not really comparable.

10.6 Conclusions

In a review of caging experiments in sedimentary habitats, Peterson (1980) argues that, since numbers and biomass always seem to increase under the cage, this is strong evidence that deposit-feeders are not normally food-limited but are maintained below the carrying capacity of the environment by other species disturbing the sediment. Once the disturbers are excluded by the cage, carrying capacity can be approached. A counter-argument is that since caging experiments are usually short-term, there is insufficient time for competitive exclusion to occur (the time-scale needed for this can be very long, because species simply reduce their growth rates and coexist without mortality increasing), but that eventually this would reduce numbers and biomass again.

But in fact the effects could be nothing more than artifacts of the cage. In Table 10.2 Reise's (1977) data are arranged in a different species order to that presented by the author, in order to emphasise the species that have increased in abundance under the cages: *Capitella capitata*, *Polydora* spp., *Peloscolex benendeni*, *Heteromastus filiformis*, *Pygospio elegans* and *Tharyx marioni*. As was discussed in Chapter 8, all of the above, with the exception perhaps of *Tharyx*, are classical opportunist species which respond to disturbance by increasing in abundance. And putting a cage over the sediment disturbs the community (sedimentation rate increases under the cage, current flow is reduced, light is less, etc., etc.). Thus the experiment not only excludes potential disturbers, but also itself has a very powerful disturbing effect, the result of which is that the classical opportunists increase in abundance. I doubt, therefore that Reise's experiment can in fact be used to argue that numbers really do increase by a factor of 20 as a consequence of predator exclusion. If one sums the numbers of individuals of all of the above species and substracts this figure from the total, then the cage/control figures are 607/150, 2282/828, 1559/477 respectively, for the different time-periods. The dramatic 20-fold increases that Reise attributes to the effect of predators are thus reduced to a consistent 3-fold increase. The major change is in the increased abundances of the bivalve molluscs, particularly *Cerastoderma edule* and *Macoma balthica* which are protected from predation.

Experiments done at depths of 23 m in the Oslofjord by John-Arthur Berge have shown that cages placed on a muddy bottom in an organically enriched area have remarkably little effect; species ranks were the same in both experiment and control. In an unpolluted area, however, the total number of individuals was significantly higher in the cage (4779) than in the control (2849). Nearly all the increased abundance under the cage was contributed by small, newly settled tellinacean molluscs, which were protected from predation. Berge concludes that in the areas of subtidal sediments that were studied, predation has relatively little effect on structuring the communities. This is in marked contrast to the intertidal data, but if my analysis above is correct, then separating cage artifact effects from the true predator exclusion experiment leads to the same conclusion for both intertidal and subtidal communities – that predators which disturb communities (*sensu* Huston, Chapter 6) play only a minor role in structuring benthic communities, since they probably only affect newly settled spat of bivalves.

If this is the case, then Peterson's argument that benthic communities are constantly held below carrying capacity by disturbance and are not food-limited, does not hold for the majority of infaunal species. There is little doubt that benthic communities in the deep sea and in the Baltic Sea *are* food limited. In these areas if densities of bivalves are increased over the normal abundance then growth rates are reduced, suggesting that indeed food is limiting for deposit-feeders and suspension-feeders. More carefully controlled manipulation experiments will solve this interesting and important controversy. We should no longer be interested in the gross all-encompassing predator exclusion experiments, but rather concentrate on more carefully designed experiments – for example altering densities of species and measuring changes in various parameters, such as growth rate, *in situ*, or altering predator density rather than excluding predators altogether, in order to measure more nearly the 'natural' impact of predators – that eliminate disturbance effects. The question of food-limitation of benthic communities in sediments has highly important consequences for the more general problem of increasing fish stocks, and will be particularly important in energy-flow models of benthic systems. Again, I suspect that certain areas will be found where communities are not food-limited and other areas where they are, and that the latter is likely to be the rule rather than the exception.

But benthic ecologists have done enough theorising for the time being and now more concrete data of a natural-history type are needed before any material advances in our understanding of factors controlling the structure of benthic communities can be made.

11

Function in benthic communities

At the beginning of this book the distinction was made between structure and function, structure being defined as fluctuations in space and time of numbers of individuals and species, and function as the type and amounts of energy flowing through the system. It is, in fact, rather difficult to draw hard-and-fast lines between the two and when one is studying biomass, or even numbers, as they change through time, then one is really measuring both structural and functional changes. So the distinction is artificial, and it is preferable to study structure and function together rather than separately. An illustration of the combined approach would be a study of the effects of pollution on marine organisms that measured effects at the biochemical level (such as variations in enzyme systems along a pollution gradient), at the physiological level (in terms, say, of scope for growth, where the growth potential is measured as an index of energy input relative to output), and at the ecological level (by, say, measuring growth rates in the field). The approaches overlap and are complimentary. The above techniques are, however, done at the individual level and I have considered here mainly populations and groups of populations of different species (communities). I will begin by discussing functional attributes of even larger units than communities, namely ecosystems.

The English botanist, Tansley, was the first to use the term ecosystem, which he defined as not only the organisms but also the complex of physical variables that make up the environment. Functional studies of ecosystems really began with Lindeman's classical paper (1942) on trophic-dynamics. Rather than regarding food merely as particulate matter, Lindeman expressed it in terms of the energy it contained, thereby enabling comparisons between different systems to be made. For example, 1 g of *Ensis* is not equivalent in food value to 1 g of *Calanus*, so the two animals cannot be compared in terms of weight; but they can be compared in terms of the energy units that each gram dry weight contains. The energy unit originally used was the calorie, but this has now been superseded by the joule (J), 1 calorie being equivalent to 4.2 joules. *Ensis* contains 14 654 J g^{-1} dry wt and *Calanus* 30 982 J g^{-1} dry wt.

Lindeman suggested that the primary source of energy for any community is sunlight, which is first fixed and stored in plant material. Plants thus constitute the first trophic level in the ecosystem. When herbivores feed on plants the energy stored in the plant protoplasm is transferred to the animal's protoplasm (i.e. to the next trophic level). But, according to the second law of thermodynamics, the efficiency of the transfer can never be 100%; much energy is 'lost', for example by respiration. When carnivores consume herbivores a similar loss of energy results. Energy, therefore, constantly flows through the system, the rates of flow being measured as J m^{-2} d^{-1}.

Lindeman measured the ecological efficiency between two trophic levels of an ecosystem as

$$I_t/I_{t-1} \times 100,$$

where I is energy intake and t is trophic level. In his study area, Cedar Bog Lake in Minnesota, USA, Lindeman found an average ecological efficiency over all trophic levels of 10% (range 5.5%–22.3%). Since his work, the value 10% has been widely accepted, although in marine planktonic systems, at least, efficiencies appear to average around 22–25%.

Fig. 11.1 illustrates diagrammatically the idealised trophic structure of a typical ecosystem. Much material passes directly to the decomposer cycle to be ultimately recycled as nutrients. No system is of course closed as shown here, and material is exported and imported by migrating species and transported by environmental factors. Many species also feed at more than one trophic level, greatly increasing the complexity of the system.

The calculation of the ecological efficiency of a complete ecosystem is an inordinately complex and time-consuming task and one that has rarely been tackled. As a first approach to functional studies many workers have concentrated on comparing how much organic material is produced by different species in the course of a year. This is a much more realistic goal than calculating ecological efficiencies, and indeed is the approach that has been adopted over a long period in studies of benthic communities.

11.1 Secondary production in benthic macrofauna

Organic material produced within a community by plants is called primary production, since green plants produce their organic matter directly from sunlight via photosynthesis. Animals have to obtain their energy by ingesting other animals or plants, and hence production by animals is called secondary production.

The earliest data on production of benthic species are those of the Dane, Boysen-Jensen (1919). In an 8-year study of the fauna of a bay in the Limfjord he estimated population numbers in April each year

Fig. 11.1. Idealised trophic structure of an ecosystem. Solid lines, flow of energy; dashed lines, flow of nutrients. Energy flows only in one direction; elements are recycled.

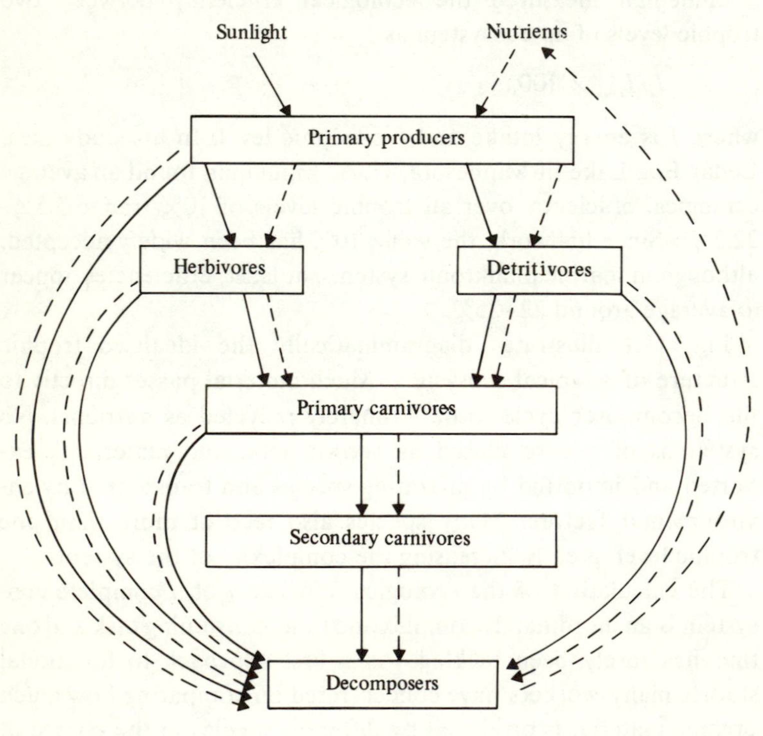

and from this and data on weights he calculated production. For example, in *Corbula gibba*, a bivalve which reproduces annually, he found in April 1912 a recruitment of 162 individuals to the population under study. These had a total weight of 3.9 g. By April 1913 there were only 90 of these individuals left, with a total weight of 3.4 g. His estimate consumption (today called elimination, E) by the plaice, *Pleuronectes platessa*, was

$$N_1 - N_2 \times 0.5\,(\overline{w}_1 + \overline{w}_2),$$

where N_1 is the original number of individuals, N_2 the final number and $\overline{w}_1$ and $\overline{w}_2$ the respective average weights. In this case

$$E = 162 - 90 \times 0.5\,(3.9/162 + 3.4/90)$$
$$= 2.23 \text{ g.}$$

He also calculated the production (P) during the year, by adding the stock (the biomass of the population at a given point in time) at the year-end (B_1) to elimination and subtracting the standing stock at the beginning of the year (B_0):

$$P = 2.23 + 3.4 - 3.9 = 1.7 \text{ g.}$$

Today this method of calculating production is still used. It can be written formally as

$$P = (B_1 - B_0) + E. \qquad (11.1)$$

Boysen-Jensen realised that the case of *Corbula*, which has infrequent recruitment, was a special case and that normally recruitment (in g) must be added to equation 11.1 to give a proper estimate of production. Recruits will, of course, be included in the change in standing stock biomass at the year-end.

The problem with Boysen-Jensen's method, of which incidentally he was aware, is that it assumes that elimination all occurs at the end of six months (0.5 × 1 year) and on the mean weight of animals. Boysen-Jensen regarded the likely errors resulting from this to be small, whereas, in fact, they can be very large. Predation, if it occurs at a constant rate, will lead to a constantly diminishing population, and the mean mortality (and hence biomass) will be reached well before six months.

The obvious refinement is to take samples more frequently. This was done by Howard Sanders, some 37 years after Boysen-Jensen's pioneering research. Sanders produced a curve of log weight of individuals against time (growth) and log numbers against time (survivorship). The population was assumed to be composed entirely of recruits that had settled just prior to the first sampling, so that the initial standing stock biomass (B_0) was assumed to be zero. Fig. 11.2 shows the results obtained for the polychaete, *Nephthys incisa.* Instead of annual measurements, observations were made at 3-month intervals, but the curves were derived using Boysen-Jensen's method. Elimination and growth could be calculated over each time-interval and figures for annual production obtained.

One curious fact that I have discovered from comparing Sanders' with Boysen-Jensen's papers is that at the end of his paper Sanders reports for the first time the production/biomass ($P/\overline{B}$ ratio, where $\overline{B}$ is the average standing stock biomass). In the above example $P/\overline{B}$ is 9.3435/4.3231 = 2.16. Boysen-Jensen gives the equation for production (what he calls growth) as:

$$P = E + B_1 \div \overline{B}.$$

This equation was faithfully reproduced by Sanders in his paper (p. 391) and he went on to compare annual production in different species by means of this ratio. Unfortunately, there is a gross error

Fig. 11.2. Survivorship and growth of *Nephthys incisa*. (Data from Sanders, 1956.)

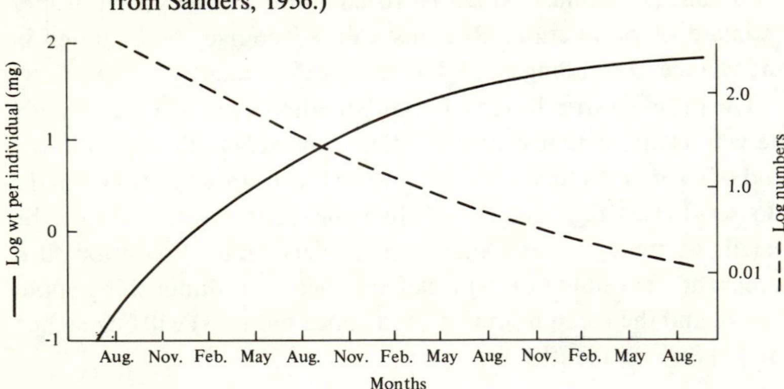

in that the '÷' sign in Scandinavian languages means minus and not divided by (although today in schools the British convention of using '−' to mean minus is taught). It seems likely, therefore, that the widely used $P/\overline{B}$ ratio arose fortuitously from a misunderstanding!

As an example of a modern production estimate I will use the data from Kirkegaard (1978) on *Nephthys hombergi*, a polychaete studied in the Isefjord, Denmark. Table 11.1 shows the detailed calculations. The mean density and biomass are estimated each month and the mean weight can be derived from these figures by division of biomass by numbers. Production is usually calculated directly rather than as the sum of growth and elimination, as Kirkegaard has done. I think that Kirkegaard's method of presenting production is easier to understand and therefore prefer this method of calculation. Traditionally production is calculated either by:

$$P(t_1 - t_2) = \sum_{t=t_1}^{t=t_2} \Delta N \, \Delta \overline{w} \qquad (11.2)$$

or

$$P = N_1 \overline{w}_1 + \sum_{t=1}^{t=2} \Delta \overline{w} \, \Delta N, \qquad (11.3)$$

where N is the number of survivors at time t, $\overline{w}$ is the average weight of an individual over time-interval $t_1 - t_2$, and ΔN and $\Delta \overline{w}$ are the changes in numbers and weight respectively over the same time-interval. Equation 11.2 calculates production increments and equation 11.3 elimination increments. Both will give the same result over similar time-periods. In the case of Kirkegaard's data, using either equation 11.2 or equation 11.3 gives the same figure for production *over the life cycle of the species*, namely 4329 mg m^{-2} (although one has to assume that the last individual is eliminated at a weight of 880 mg to balance the equation!).

Production estimates are fairly easy to obtain provided that recruitment occurs at one point in time or that age classes are easily separable. If recruitment occurs continuously at the same rate as mortality, then using the methods outlined above there would be no change in numbers, mean weight or standing stock biomass, and, hence, no production. But if the age classes can be separated then production can still be calculated. In the case of *Nephthys*, Kirke-

Table 11.1. *Production calculations for the polychaete* Nephthys hombergi *from Isefjord, Denmark, 1970–1*

Age (months)	Mean density (m^{-2})	Mean weight (mg)	Biomass ($mg\ m^{-2}$)	Growth increment[a] ($mg\ m^{-2}$)	Elimination loss[b] ($mg\ m^{-2}$)	Production ($mg\ m^{-2}$ per month)	Annual net production ($mg\ m^{-2}$)	Annual mean biomass ($mg\ m^{-2}$)	$P/\bar{B}$
0 (1 Aug.)	47	—	—	—	—	—			
1	42	20	840	840	50	890			
2	38	40	1520	760	40	800			
3	34	48	1632	272	16	288			
4	31	50	1550	62	3	65			
5	28	51	1428	28	2	30			
6	25	52	1300	25	2	27			
7	22	53	1166	22	2	24			
8	20	54	1080	20	1	21			
9	18	55	990	18	1	19			
10	16	60	960	80	5	85			
11	15	80	1200	300	10	310			
12	14	120	1480	560	20	580	3139	1262	
13	13	160	2080	520	20	540			
14	10	185	1850	250	38	288			
15	8	200	1600	120	15	135			
16	6	208	1248	48	8	56			
17	5	209	1045	5	1	6			

18	3	210	630	3	1	4			
19	3	212	636	6	0	6			
20	2	214	428	4	1	5			
21	2	216	432	4	0	4			
22	1	220	220	4	2	6			
23	1	245	245	25	0	25			
24	1	280	280	35	0	35	891	1110	
25	1	320	320	40	0	40			
26	1	340	340	20	0	20			
27	1	355	355	15	0	15			
28	1	360	360	5	0	5	80	344	
Total							4329	2497	1.7

From Kirkegaard (1978).

[a] The growth increment is calculated as the number of animals at time t_1 multiplied by the increase in mean weight over the time period t_0 to t_1:

$$= N_1(\bar{w}_1 - \bar{w}_0),$$

where N is the number of individuals and $\bar{w}$ is the mean weight.

e.g. 42(20–0) = 840; 38(40–20) = 760; 34(48–40) = 272; etc.

[b] Elimination loss is calculated as the difference in mean density between one time-period and the next multiplied by the average mean weight over the time-period:

$$= N_1 - N_0 \times 0.5(\bar{w}_1 - \bar{w}_0).$$

e.g. 47–42 × 0.5(20–0) = 50; 42–38 × 0.5(40–20) = 40; 38–34 × 0.5(48–40) = 16; etc.

gaard was able to use the chitinous jaws to age the specimens, since they bear discernible growth marks. Fig. 11.3 shows fairly typical size-frequency data for two species of bivalve, *Mya arenaria* and *Macoma balthica*. Here the age classes were separated by growth rings and by using probability paper. The latter method is widely used and details can be found in Crisp (1971). Production was calculated from the curves in Fig. 11.3, as shown in Table 11.2.

Kirkegaard calculated his $P/\overline{B}$ ratio over the whole life-span of the species whereas Sanders did not, and thus Sanders underestimated the true ratio.

Production estimates are most difficult when age classes are not easy to separate. In this case one needs to know the density, mean weight and specific growth rate of each size-class. A recent example of the calculation of production in such a case is the data on the polychaete *Nereis diversicolor* (Heip & Herman, 1979). Here the growth rate was calculated by measuring the time taken for an individual to grow from one size class to the next. This can be done by plotting the data for two consecutive time-intervals (here fortnightly intervals) and sliding the *x*-axis of one along the *x*-axis of the other until the patterns have the maximal fit. Heip & Herman, however, used a statistical technique involving the calculation of cross-correlation functions; their paper should be consulted for more details. Fig. 11.4(*a*) shows the data, which indicate that the smallest worms grew from one class to the next slowly at first and then more rapidly, but that once over 12 mm growth rate decreased linearly. Fig. 11.4(*b*) shows weight-specific growth rate determined from (*a*). From these data production was estimated at 398 g wet wt $m^{-2}yr^{-1}$ or 61.3 g dry wt $m^{-2}yr^{-1}$. The average biomass was 158 g wet wt m^{-2}; the $P/\overline{B}$ ratio was 2.51.

11.2 Production estimates in meiofauna

There are within the meiofauna a fairly large number of species where reproduction occurs continuously and asynchronously and where cohorts (distinct size classes) cannot be separated. It is extremely difficult to calculate production for such species using the traditional methods, and one often has to resort to indirect methods. For example, from field samples of the meiobenthos off Northumberland it was impossible to ascertain either

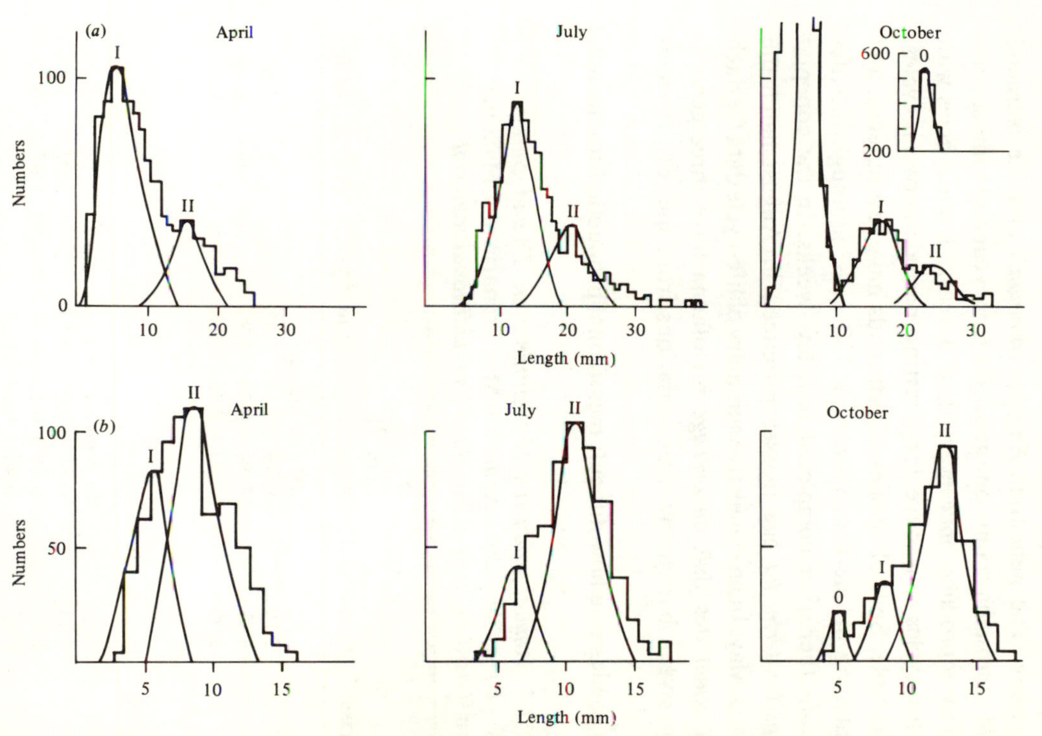

Fig. 11.3. Size-frequency histograms for (*a*) *Mya arenaria*, (*b*) *Macoma balthica* from eastern Canada. Normal curves for each age class (I and II) were obtained by probability-paper analysis and interpretation of growth rings. (After Burke & Mann, 1974.)

the number of nematode generations per year or the number of eggs laid. Nematodes were therefore brought into the laboratory in order to obtain estimates of generation times.

It has been found, however, that generation times of meiofauna vary widely. The nematode *Enoplus communis* and the ostracod *Cyprideis torosa* have only one generation per year and some species may only reproduce once every 2 or 3 years. Even species with short life-cycles may have long resting periods between cycles. For example, Several species of ostracods have resting eggs, and egg hatching in one nematode took 150 days, although the life-cycle was thereafter completed in a few weeks. In the copepod *Asellopsis intermedia* egg masses produced in August hatch the following May. In an exhaustive summary of life-cycle data Gerlach (1971) concludes that *on average* meiofauna have three generations per year, but clearly large variations from species to species occur.

Just to illustrate how difficult meiofaunal production studies are,

Table 11.2. *Annual production (P), elimination (E) and average biomass ($\bar{B}$) (as flesh dry weight) for* Mya arenaria *and* Macoma balthica, *based on data in Fig. 11.3 and additional material*

Year class	P (g m^{-2}yr^{-1})	$\bar{B}$ (g m^{-2}) ±S.E.	E (g m^{-2} yr^{-1})	$P/\bar{B}$	$E/\bar{B}$
		Mya arenaria			
0	2.19	0.79 ± 0.14	1.45	2.77	1.84
I	6.42	2.58 ± 0.28	3.83	2.49	1.48
II (measurable only 5 months)	3.00	1.20 ± 0.46	3.88		
		Macoma balthica			
0	0.23	0.11 ± 0.03	0.16	2.09	1.45
I	1.15	0.69 ± 0.13	0.51	1.67	0.73
II (measurable only 5 months)	0.55	0.46 ± 0.17	1.39		

From Burke & Mann (1974).

the life-cycle of *Chromadorita tenuis* is illustrated (Fig. 11.5). This was obtained from laboratory studies. Two adults produce 20 fertilised eggs which, assuming they hatch after 5–6 days, reach the adult stage on day 21. We have no reliable estimate of elimination, but assuming this is around 10% per day during juvenile life, we can obtain an average standing stock biomass ($\overline{B}$) of 3.2 μg wet weight, with elimination of 20 animals totalling 9.8 μg during the life-cycle. Thus, the life-cycle turn-over rate is 9.8/3.2 = 3. Assuming that there are three generations per year, the annual $P/\overline{B}$ ratio is 9. However, the relation between P and $\overline{B}$ over time depends on the shape of the growth and elimination curves, and methods such as those used here must be treated with great caution.

Fig. 11.4. Growth of *Nereis diversicolor*. (*a*) Time taken for different size classes to grow to the next size class. (*b*) Relationship between weight-specific growth rate (G_i) and size:

$$G_i = \frac{\ln w_{i,2} - \ln w_{i,1}}{t_2 - t_1},$$

where $w_{i,1}$ and $w_{i,2}$ are weights of an individual in size class i, and t_1 and t_2 are two time-intervals. (After Heip & Herman, 1979.)

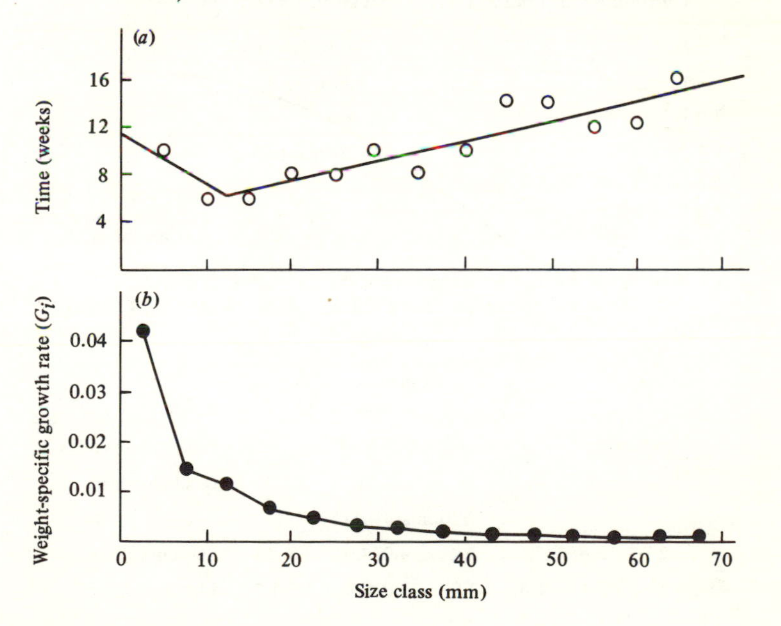

Heip (1976) has recently produced an ingenious alternative solution to the problem of estimating $P/\overline{B}$ ratios for species that reproduce continuously. Using a combination of field estimates of population density and laboratory studies of population growth rates, he calculates numbers eliminated from the population. The details of his method are too complex to reproduce here, but in essence he calculates the $P/\overline{B}$ ratio from numbers of individuals, since the mean weight appears in both numerator and denominator and can therefore be cancelled out. In a population of the brackish-water copepod *Tachidius discipes* he found 4175 individuals eliminated per 100 cm^2 and a mean standing stock biomass of 278 individuals per 100 cm^2; this gives a $P/\overline{B}$ ratio of 4175/278 = 15.0, which is closely similar to that obtained by more complex methods based on calculations of successive elimination increments. This method looks to be highly promising for estimating production in other meiofaunal species, but very few species have, as yet been studied. The figure of 15 obtained for *Tachidius* is similar to that for planktonic copepods that have many generations per year; $P/\overline{B}$

Fig. 11.5. Calculation of turn-over rate in the nematode *Chromadorita tenuis*. (After Gerlach, 1972.)

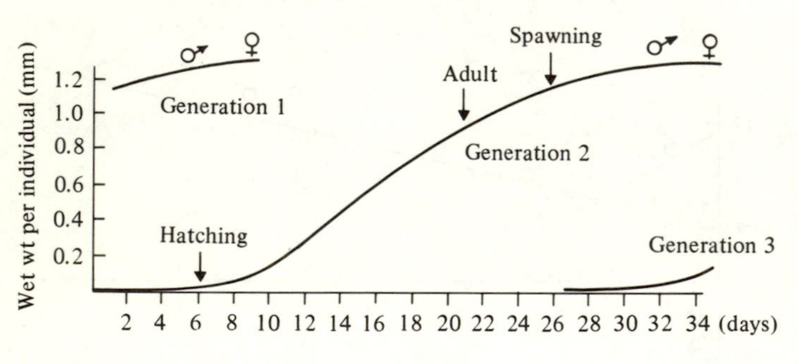

				Numbers				
	2	1			4	3	2	Adult
20	18	13	8	5				Eggs and juveniles
	2	6	6	3	1	1	1	Dead animals
				Biomass (μg)				
2.9	3.0	2.6	3.2	4.0	4.0	3.3	2.4	Standing stock
	0.1	1.8	2.8	2.0	0.9	1.0	1.2	Elimination

ratios will be appreciably lower in species such as *Enoplus* and *Cyprideis* that only have one generation a year.

From these limited and scanty data it is difficult to draw many meaningful conclusions on meiofaunal production figures. They are likely to be as varied or maybe more varied than those for macrofauna.

11.3 Production : biomass ratios

The idea behind the calculation of the $P/\overline{B}$ ratio is that it allows populations of different biomass to be compared on a common basis. Table 11.3 illustrates published data on $P/\overline{B}$ ratios for a variety of macrobenthic species. In a number of cases $P/\overline{B}$ ratios have been calculated for the same species at different localities. In some of these the values are in good agreement (e.g. *Nephthys hombergi*, 1.9 and 1.7), whereas others show a wide discrepancy (e.g. *Macoma balthica*, 2.1 to 0.8). It is quite possible that methodological differences account for much of the difference, since the lower figure may, for example, have underestimated the growth of the smallest juveniles.

One of the attractions of calculating $P/\overline{B}$ ratios was the hope that a common pattern would emerge so that production could be calculated merely from estimates of biomass. However, there are many factors that influence $P/\overline{B}$ ratios. Low temperatures, slow growth rates, and varying predation rates all lower the ratio. But both growth-rate changes and predation alter the ratio by causing changes in the age distribution of species, and it is population growth rate and size structure that are the major factors influencing $P/\overline{B}$ ratios. Recently Robertson (1979) plotted most of the data in Table 11.3 against life-span in years (Fig. 11.6). The equation representing the data was

$$\log_{10} P/\overline{B} = 0.660\ (\pm\ 0.089) - 0.726\ (\pm\ 0.147) \log_{10} L,$$

where L is life-span in years. All groups (bivalves, gastropods, polychaetes, crustaceans and echinoderms) fitted the above equation ($r = -0.8350$) and there was no significant difference between each group plotted separately. Thus, a significant and important trend has emerged: that there is a general relationship between annual $P/\overline{B}$ values and life-span for all marine macrobenthos. Robertson suggests that the equation could be used to predict production from biomass, provided that the sampling gear is not age-selective and that age can be determined. As yet there are insufficient data on meiofauna to be able to see whether the same equation applies or whether a different equation will be appropriate.

$P/\bar{B}$ ratios are often all that is studied in relation to the energy

Table 11.3. *$P/\bar{B}$ ratios of benthic macrofauna*

Polychaeta		Bivalvia	
Ampharete acutifrons	5.5	*Pillucina neglecta*	4.0
Ampharete acutifrons (2)[a]	4.6	*Tagelus divisus*	3.9
Pectinaria hyperborea	4.3	*Macoma incongrua*	3.7
P. californiensis	4.3	*Veremolpa micra* (2)	3.2
P. koreni	3.1	*Theora lubrica* (2)	2.9
Harmothoe imbricata	2.6	*Cerastoderma edule*	2.9
Nereis diversicolor	2.5	*Dosinia elegans*	2.8
Terebellides stroemi	2.3	*Mya arenaria*	2.5
Nephthys incisa	2.2	*Tellina martinicensis*	2.4
Ammotrypane aulogaster	2.1	*Yoldia limatula*	2.3
Cistenoides gouldii	1.9	*Macoma balthica*	2.1
Nephthys hombergi	1.9	*Pandora gouldii*	2.0
Nereis diversicolor (2)	1.8	*Crassostrea virginica*	2.0
N. hombergi (2)	1.7	*Musculus senhausius*	1.7
Neanthes virens	1.6	*Macoma balthica* (2)	1.5
Spiophanes kroyeri	1.4	*Tellina deltoides*	1.4
Lumbrinereis fragilis	1.3	*Abra nitida*	1.1
Chaetozone setosa	1.3	*Venerupis aurea*	1.1
Heteromastus filiformis	1.0	*Mytilus edulis*	1.0
Arenicola marina	1.0	*Macoma balthica* (3)	0.9
Glycera rouxi	0.4	*Chione cancellata*	0.8
Nepthys australensis	0.4	*Macoma balthica* (4)	0.8
Crustacea		*Scrobicularia plana*	0.5
Calianassa australiensis	3.9	*Mya arenaria* (2)	0.5
Crangon septemspinosa	3.8	*Mercenaria mercenaria*	0.3
Neomysis americana	3.7	*Venerupis pullastra*	0.2
Ampelisca brevicornis	3.5	*Scrobicularia plana* (2)	0.2
Macrophthalamus latifrons	3.2	*Cerastoderma edule* (2)	0.2
Palaemonetes pugio	2.8	Echinodermata	
Alpheus euphrosyne	2.7	*Asterias forbesii*	2.6
Calocaris macandrae	0.1	*Moira atropes*	1.0
Veremolpa micra	4.9	*Brissopsis lyrifera*	0.3
Theora lubrica	4.1		

[a] Figures in parentheses indicate number of determinations.

Fig. 11.6. Annual $P/\bar{B}$ ratio against life-span for marine macrobenthos. (After Robertson, 1979.)

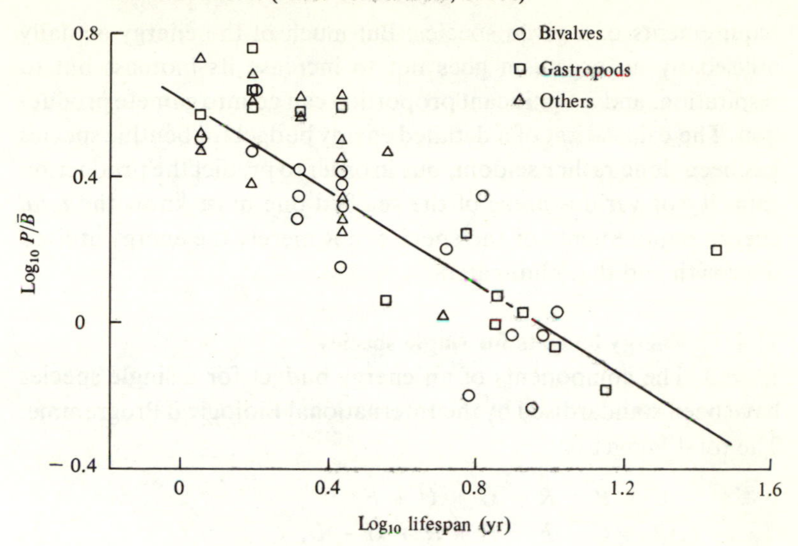

Fig. 11.7. Oxygen consumption of *Scrobicularia plana* in a continuous-flow respirometer. (After Hughes, 1970.)

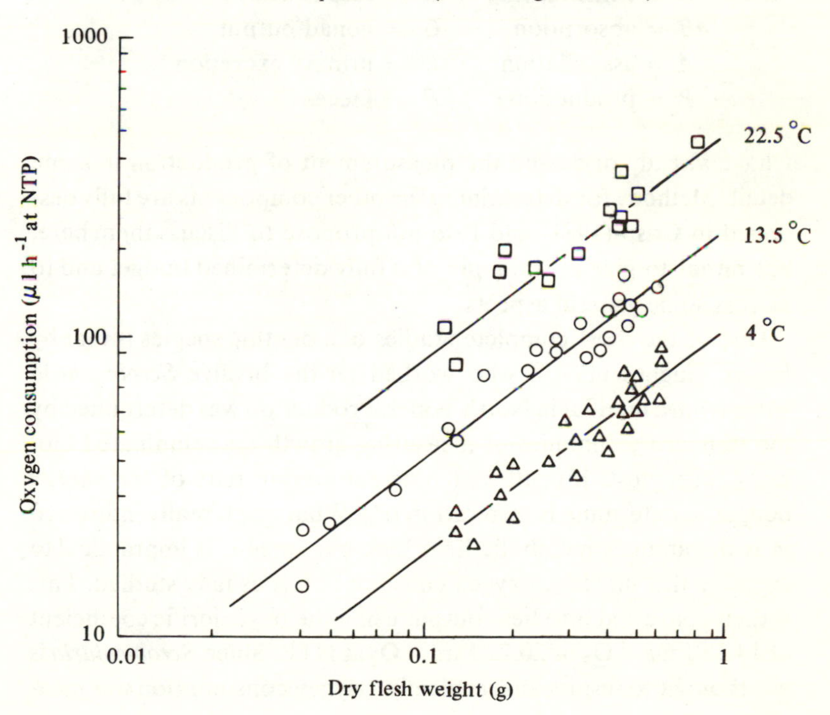

requirements of a given species. But much of the energy actually utilised by an organism goes not to increase its biomass but to respiration, and a significant proportion can go into gamete production. The calculation of a detailed energy budget for benthic species has been done rather seldom, but in order to predict the production capacity of various areas of the sea-bed one must know the *total* energy requirements of the species, not merely the energy utilised in growth and that eliminated.

11.4 Energy budgets for single species

The components of an energy budget for a single species have been standardised by the International Biological Programme. The total budget is:

$$C = P + R + G + U + F,$$
$$AB = C - F = P + R + G + U,$$
$$A = P + R + G,$$

where C = consumption, R = respiration
AB = absorption, G = gonad output
A = assimilation, U = urinary excretion
P = production, F = faeces

I have already discussed the measurement of production in some detail. Methods for determining the other components are fully described in Crisp (1971) and I do not propose to discuss them here, but rather to give an example of a fully determined budget and to discuss some general aspects.

One of the most complete studies of a benthic species is that by Roger Hughes (1970), who worked on the bivalve *Scrobicularia plana* from a mudflat in North Wales. Production was determined by the standard techniques of measuring growth and eliminated biomass. The most difficult and time-consuming part of the energy budget to determine is respiration (R). What one is really interested in is the animals metabolic heat loss, but since it is impractical to measure this directly, oxygen consumption is usually studied. This is then converted into heat output using the oxycalorific coefficient of 14.15 J mg^{-1} O_2 or 20.22 J ml^{-1} O_2 at NTP. Since *Scrobicularia* is not thought to respire anaerobically, oxygen consumption was measured using a continuous-flow respirometer. This is basically a sealed vessel with a constant flow of water through it into which the animal is put. The difference in oxygen content between the inflow and outflow is multiplied by the current speed to give the rate of oxygen consumption. The vessels were maintained within the range of temperatures found in the field. Typically there is a wide scatter in the values obtained. Fig. 11.7 shows Hughes' data. For each of the temperatures used a regression equation was calculated and using field data on temperatures the metabolic heat loss could be estimated using the oxycalorific coefficient.

Determining the amount of energy that goes into gamete production was difficult because it was not possible to induce the animals to spawn in the laboratory. (Had this been possible it would have been easy to collect the spawn and measure its energy content directly.) The method adopted was to determine the fall in dry flesh weight that occurred after spawning and multiply that by the energy content per gram. Fig. 11.8 shows the data. Urinary wastes could not be estimated and in the marine environment this component of the energy budget is frequently neglected. Faecal pellets were collected from animals of various sizes and over the

Fig. 11.8. Seasonal variation in dry flesh weight of *Scrobicularia plana* with a shell length of 40 mm. ○, 1966; △, 1967; ▲, computed from regression lines with a fixed slope of 3.0. (After Hughes, 1970.)

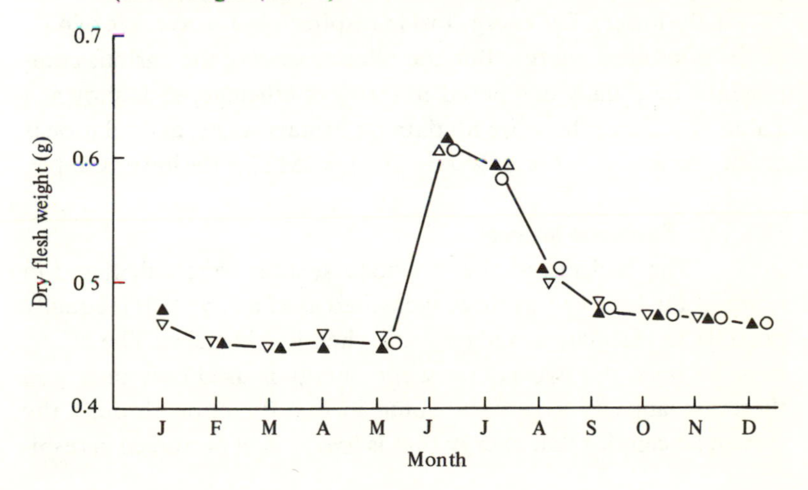

annual temperature range. From these data the amount of energy lost in the faeces for each month for the total population could be calculated. The energy content of faecal pellets did not in fact vary much during the year.

Hughes studied two populations, one high up the shore and one lower down; the two separate energy budgets and the average combined budget were (in kJ m^{-2}):

Lower sample: $C = P + R + G + U + F$
$C = 519 + 1993 + 268 + \text{n.d.} + 1624$
$C = 4404$

Upper sample: $C = 73 + 228 + 17.6 + \text{n.d.} + 198$
$C = 516.6$

Average: $C = 296 + 1110 + 143 + \text{n.d.} + 911$
$C = 2460$

As a check on the budget it was possible to calculate consumption also by determining the energy content of the sediment that *Scrobicularia* ingested, making a correction for pseudofaeces produced, and then subtracting the energy lost as faeces. The figures in this case were (in kJ m^{-2}): lower, 3562; upper, 436; average, 1999. The differences between these and the first set of figures were 20, 16 and 19% respectively.

A surprisingly large proportion of the energy budget goes to gamete production (31% of the production in the upper sample and 52% in the lower). The energy lost in respiration is an average of 45% of the consumed energy. But the relative sizes of the various components are usually compared in terms of efficiencies, as shown in Table 11.4. Since there are no data on urinary waste, assimilation is defined as $A = C - F$, e.g. $4404 - 1624 = 2512$ for the lower sample.

11.5 Elemental budgets

The budgets in the previous section were calculated in terms of the total energy flow, measured as kJ $m^{-2} yr^{-1}$. It is equally possible to calculate a budget for a chemical element. The difference between the two is that while energy is used only once and therefore appears as a flow going in one direction through the system (except for that energy that is lost as heat produced in respiration), elements can be recycled. This difference was illustrated in Fig. 11.1.

Most of the organic matter in ecosystems is usually found as detritus. This detritus is then broken down by micro-organisms and the elements can then be recycled. Carbon has been widely used in element budgets but there are advantages in using nitrogen. Nitrogen values can be used to measure excreted waste and furthermore nitrogen is often the nutrient that limits primary production in the sea. By calculation of a nitrogen budget for a given species or for the whole ecosystem more information on factors directly relevant to the functioning of the system are obtained.

Specific elements are assimilated by animals with a greater efficiency than is organic matter as a whole, so element budgets will give different efficiencies compared with energy budgets. To obtain the best picture of the trophic relations of a species both energy and element budgets should be compiled, so that in addition to knowing the total energy flow the carbon budget could be used to estimate respiration (as carbon dioxide produced) and the nitrogen budget to estimate growth (as protein anabolism) and excretion. But this has not been done for any species and would be inordinately time-consuming.

Element budgets are usually for carbon, nitrogen, phosphorus or sulphur. I will take as an example the carbon budget for a member of the benthic meiofauna, the harpacticoid copepod *Asellopsis intermedia*. It was studied at Firemore beach on the west coast of Scotland by Lasker, Wells & McIntyre (1970). Growth was estimated

Table 11.4. *Efficiencies of various parts of energy budget for* Scrobicularia plana

	Assimilation (gross production) (kJ m^{-2} yr^{-1})	Production (kJ m^{-2} yr^{-1})	Respiration efficiency ($R \times 100/A$)	Assimilation efficiency ($A \times 100/C$)	Net growth efficiency ($P \times 100/A$)
Lower	2780	519	72	63	22.9
Upper	318.6	73	72	62	18.7
Average	1549.3	296	72	63	19.1

Data from Hughes (1970).

using the traditional methods of size-frequency histograms and dry weights, and organic carbon content obtained by standard methods. Quite an appreciable amount of carbon can be 'lost' as moults, so the carbon content of the moults was also measured. Since the animal is rather small (0.6 mm long) special techniques had to be employed to measure the respiration rate, and a highly sensitive ultramicro-respirometer was used. Again no estimate of excretion was obtained. Table 11.5 shows the results, and it can be seen that remarkably little energy goes to production.

If the nitrogen rather than carbon budget is determined then excretion can be estimated. Harris (1973) studied the nitrogen budget for another harpacticoid copepod, *Tigriopus brevicornis*, which lives in rock-pools. In order to estimate excretion large numbers of individuals from each life stage were separated and placed in beakers after they had been allowed to feed for a number of hours. The amounts of nitrogen excreted over 4–10-hour periods in the dark were then measured directly on water samples from the beakers. It was found that over 75% of the nitrogen was excreted as ammonia and that up to 30% of the body nitrogen was excreted per day (an average of 27.2 $\mu g\ N\ mg^{-1}$ dry wt). From this example it is clear that excretion can indeed be a substantial source of energy loss and should not necessarily be ignored as a relatively trivial part of energy budget calculations. Fig. 11.9 shows the data on nitrogen in the body and excretion over time. By subtracting the initial from

Table 11.5. *Elemental budgets for the harpacticoid copepods* Asellopsis intermedia *(as carbon)*[a] *and* Tigriopus brevicornis *(as nitrogen)*[b]

	Percentage of element assimilated			
	Production	Respiration	Gametes	Moults
A. intermedia				
Male	7	90	0.2	2.8
Female	5.3	81.6	11.4	1.7
T. brevicornis	3.9	72.9	22.6	0.4

[a] Data from Lasker *et al.* (1970).
[b] Data from Harris (1973).

the terminal figure for body nitrogen, growth was estimated as using 1.1829 μg N. Harris further calculated the nitrogen required for what he termed 'metabolism', which he defined as the area under the curve of nitrogen excreted in Fig. 11.9; this gave a value of 5.5 μg N.

Table 11.5 summarises the two element budgets in terms of the percentage of the total amount of the element ingested in the food that is assimilated, and they show remarkably good agreement. Such discrepancies as there are may reflect either species differences (for example more energy is used in respiration in the smaller *Asellopsis*) or methodological differences.

I know of no other elemental budgets for sediment-living species. A phosphorus budget has been calculated for the bivalve *Modiolus dimissus*, but that species is not a true sediment-living organism. So here is a potentially fruitful area of research.

11.6 Community metabolism

It would be an inordinately complex task to obtain an energy budget for every single member of a given community. The alternative, therefore, is to take a holistic approach and mea-

Fig. 11.9. Growth and excretion in terms of nitrogen in *Tigriopus brevicornis*. (Modified from Harris, 1973.)

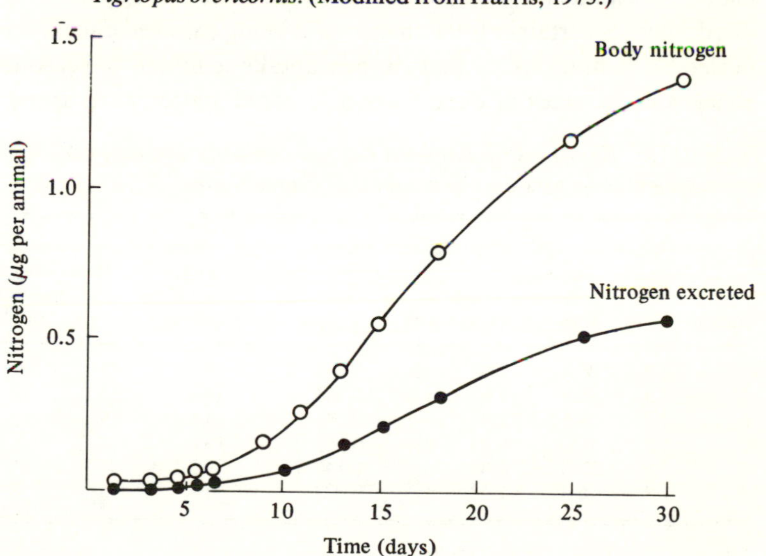

sure the metabolism of the community as a whole. If the community in question is strictly aerobic, then there is no great methodological problem, as metabolism can be measured in terms of oxygen uptake.

11.6.1 Oxygen uptake

Teal & Kanwisher (1961) suggested that the rate of oxygen uptake by the sediment surface could be an integrated measure of both aerobic and anaerobic metabolism. Measurements have been made by placing black bell-jars over the sediment and by means of an electrode system monitoring the oxygen uptake directly. The technique has been used intertidally and also down to depths of 180 m, using TV cameras to orientate the positioning of the jars. Table 11.6 shows the results of Pamatmat & Banse (1969) from the sea-bed in Puget Sound, USA. There was no relation between oxygen consumption and mean grain size or silt–clay fraction, nor with organic content or organic nitrogen, nor with ash-free dry weight of the macrofauna. Temperature did, however, have an influence, higher temperatures giving higher oxygen consumptions. The lack of correlation between oxygen consumption and organic matter seems to be a common phenomenon. Much of the organic matter in sediments is presumed, therefore, not to be easily oxidised; there is certainly little chance of it being oxidised once it gets buried. It is more likely that oxygen uptake is linked to seasonal changes in the rates of deposition of organic matter, such as sedi-

Table 11.6. *Relationship between oxygen consumption and various environmental factors on the sea-bed of Puget Sound, USA*

Station	Mean grain size (phi)	Percentage silt–clay	Organic matter: Total (% dry wt)	Organic matter: Nitrogen (% dry wt)	Macro fauna (ash-free dry wt)	Mean oxygen consumption (ml $m^{-2}h^{-1}$)
4	4.0	36.9	2.78	0.059	10.6	33
5	4.3	42.8	3.92	0.083	31.4	25
6	3.4	14.6	2.45	0.046	4.6	19
7	1.8	70.0	6.87	0.162	19.6	17
10	5.6	72.3	6.45	0.148	9.3	35
11	1.2	6.3	1.43	0.023	6.6	35

Data from Pamatmat & Banse (1969).

menting particles from the plankton. Davies (1975) found that in a Scottish sea loch the oxygen consumption was equivalent to the amount of carbon settling.

The influence of the fauna on oxygen consumption is surprisingly small. One technique used to estimate the relative importance of bacterial and faunal oxygen uptake is to measure the uptake under a bell-jar before and after the injection of antibiotics into the jar. The antibiotics are presumed to inhibit the bacteria without affecting other organisms, and thus the difference in oxygen consumption before and after the antibiotic injection gives an estimate of bacterial respiration. In practice inhibition is seldom more than 90%, and in some cases oxygen consumption can increase as a result of one species of bacterium not being killed and so being able to increase in abundance because of the lack of competition. After bacterial respiration has been measured, the chemical oxygen demand can be estimated by injecting formalin to kill the fauna.

Working on the fauna of a carbonated coral beach in Bermuda, Smith, Burns & Teal (1972) determined the respiration rates of the common macrofauna in the laboratory, and using estimates of meiofaunal and microfaunal respiration they were able to compartmentalise the community respiration (Table 11.7). It is quite possible, however, that in the field many organisms will respire at much higher rates than is typical in laboratory experiments done

Table 11.7. *Partitioning (%) of community metabolism estimated by bell-jars placed* in situ

(*a*) *Coral sand Bermuda*	
Bacteria	35.5
Microflora–microfauna	60.2
Meiofauna	1.6
Macrofauna	2.6
(*b*) *Intertidal mudflat, Washington*	
Microflora	47
Bacteria/microfauna plus meiofauna	33
Macrofauna	20

Data for May, from (*a*) Smith *et al.* (1972), and (*b*) Pamatmat (1968).

in highly oxygenated water. Thus, the use of laboratory data to interpret *in situ* community metabolism data requires extreme caution. Smith *et al.* found no chemical oxygen demand. The coral sand studied in Bermuda is rather atypical, however, in that pore size is large, the sediments are highly oxygenated, and bacterial respiration is unusually low. Pamatmat (1968) estimated for a mudflat in Washington, USA, not only the partitioning of community respiration between microfauna and bacteria/microfauna plus meiofauna, but also the contribution of the benthic algae (microflora). He used dark and light bottles to estimate gross production of the microflora and then estimated the oxygen consumed. The partitioning of the oxygen demand (Table 11.7) shows that bacterial respiration was much higher than that of the macrofauna. Seldom, in fact, does macrofaunal respiration contribute more than 20% of the community respiration and bacteria usually account for 50–80% of the total. The role of the meiofauna and microfauna is certainly not constant and it is likely that, in general, they consume as much oxygen as do the macrofauna. In sediments, however, there are many species that are facultative anaerobes and most of the bacteria concerned with the sulphur cycle are obligative anaerobes. In anaerobic respiration the terminal electron acceptor used may be H_2S, CH_4, NH_4^+, Fe^{2+}, NO_3^-, etc. Thus, in most sediments the data obtained from oxygen uptake studies may be underestimates of community metabolism, since the anaerobic pathways are neglected.

The ideal method of measuring community metabolism would be to use the heat energy produced, since heat is a product of both aerobic and anaerobic respiration. Direct calorimetry has been tried on marine sediments, but the techniques are extremely difficult, as is interpretation of the results obtained, so that in fact chemical methods (often based on oxygen uptake) are preferred.

11.6.2 *Biochemical methods of measuring community metabolism*

All organisms – aerobes, anaerobes and fermenters – are similar in that their metabolism results in the production not only of heat but also of hydrogen and electrons (by dehydrogenation reactions). Except in the case of fermentative bacteria (which are thought not to have an electron transfer system) the hydrogen and electrons are transferred to the terminal acceptor in the electron transfer chain. Thus the total community metabolism of macrofauna, meiofauna, microfauna and bacteria can be measured in terms of the rate of transfer of hydrogen or electrons to these acceptors, which in turn can be estimated by the activity of the enzyme catalyst for such transfer, hydrogenase. The method traditionally used is, in general terms, to measure the rate of transfer of electrons to an artificial electron acceptor. Dehydrogenase activity is usually measured by the rate of reduction of 2,3,5-triphenyltetrazolium chloride (TTC) to formazan. The enzyme is extracted and solubilised and its initial and final activity measured on addition of a superabundance of substrate (i.e. electron acceptor). Thus, the technique probably indicates the metabolic *potential*, which may not necessarily be the metabolic rate *in situ*.

The sort of results obtained are illustrated in Fig. 11.10, which shows dehydrogenase activity in cores of sediment taken from an intertidal beach in Bermuda. Further studies showed that between 80 and 90% of the activity was associated with the surface of sand grains, which was interpreted as it being due to bacterial and microfloral and microfaunal activity. This agrees well with the data from Smith *et al*. (Table 11.7), also from Bermuda, where over 90% of the metabolism was held to be due to these groups.

Measurement of dehydrogenase activity is a simpler technique

Fig. 11.10. Vertical distribution of dehydrogenase activity (measured as μmol formazan reduced h^{-1} g^{-1} dry sediment) in cores from a sand beach in Bermuda. (Data from Wieser & Zech, 1976.)

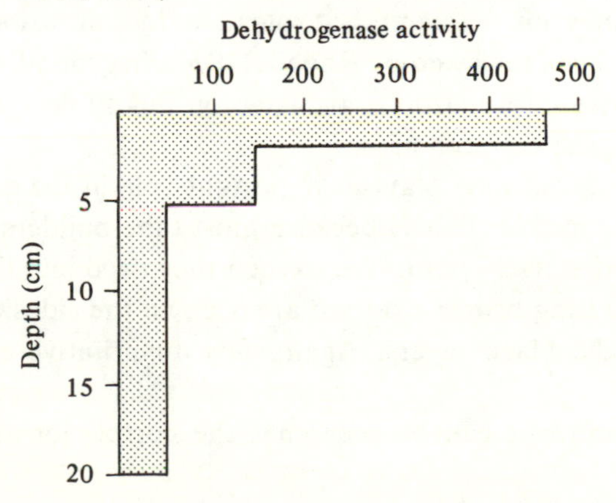

than *in situ* oxygen consumption and looks, therefore, to be a highly promising tool. Also, it is very difficult to estimate accurately the biomass and numbers of bacteria and microfauna, so the biochemical methods would seem to offer many advantages.

11.7 Factors controlling community metabolism

In the previous section it was mentioned that total oxygen uptake was found not to be correlated with the total organic content of the sediment but rather with the oxidisable fraction. This is because much organic matter becomes buried within the sediment and often large pools of undegraded organic matter occur. Within the buried sediment anaerobic metabolism produces sulphides, which give the sediment a black colour often accompanied by a smell of hydrogen sulphide. In areas where eutrophication is a problem, such as the inner part of the Oslofjord, the superabundance of organic matter settling out in the water column uses all the available oxygen and leads to anaerobic conditions where black sediment occurs at the surface. Such sediments are usually devoid of life.

Normally, however the sediment has a brown surface layer where aerobic conditions hold and the anaerobic layer is found beneath this, the exact depth depending on grain size, water flow etc. (see Chapter 2). The extent of this brown layer has been claimed as being limited by the rate of oxygen diffusion into the sediment, in which case the rate of diffusion, and thus of course the temperature, will affect the rate of oxidation of settling organic matter. I suspect, however, that the reworking activities characteristic of most deposit-feeders, during which sediment is brought up to the surface from depths, often play a more significant role than does diffusion. Such activities have been reviewed by Rhoads (1974). In general, though, little quantitative information is available on this subject and it is a highly promising research area.

Tube-builders can also play an important role in the degradation of organic matter. This is because most tube-builders respire aerobically and actively pump the oxygen they need into the sediment, often causing brown oxidised areas down the sides of their tubes within the black layers. Again, few quantitative data are available.

From the above it can be seen that the correlation that has

been found between oxygen uptake of the sediment and temperature is a complex one, related not only to the changes in diffusion rates that occur with temperature but also to the increased biological activity of reworking and pumping that raised temperatures produce.

Oxygen consumption by benthic communities seems to be proportional to the supply of sedimenting material from the photic zone. From Pamatmat's studies in Puget Sound (1968) it can be seen that there is a clear relationship between benthic respiration and chemical oxidation and phytoplankton production. This is confirmed by Davies' (1975) findings that benthic respiration increased with the amount of sedimenting organic carbon. This relationship has been formalised by Hargrave (1978) as:

$$C_o = 55\left(\frac{C_s}{Z_m}\right)^{0.39},$$

where C_o is the benthic oxygen consumption ($l\ O_2\ m^{-2}\ yr^{-1}$), Z_m is the mixed layer depth (depth of thermocline (in m) during stratification), and C_s is the annual primary production ($g\ C\ m^{-2}\ yr^{-1}$). Hargrave treated some 13 different data-sets to arrive at the above equation.

Deep-sea sediments consume oxygen at rates one or two orders of magnitude lower than those of the coastal waters described by Hargrave's equation. Also, it is generally believed that the organic matter arriving at deep-sea sediments from the plankton has been completely mineralised during its journey through the water column, and thus no clear relationship of the type above would be expected.

At high levels of primary production proportionately less oxygen is used at the sediment surface, since much material becomes buried and anaerobic metabolism takes over. The fact that in the sea loch that Davies studied all of the sedimenting carbon was utilised, probably indicates that the material, in this case sedimenting phytoplankton, was readily oxidisable. In areas where much terrigenous material occurs it has been found that as little as 23% of the available organic matter is oxidised. Thus the source of the sedimenting material plays a large role in determining how much is metabolised aerobically. On the other hand, the amount of sedimenting organic matter may not always be sufficient to meet the oxygen uptake requirements of the sediment. In such areas it is likely that organic matter produced by benthic algae *in situ* makes up the deficit. This was the case in Pamatmat's data for the intertidal mudflat in Washington. In salt marshes community oxygen consumption may be only one-half of the organic matter produced in the marsh sediments. Thus, the metabolic activity in the sediment is closely linked to the amount of sedimenting organic matter, especially that from the plankton. The interrelations of the benthos with the water column will be explored in the final chapter, which deals with models of ecosystems.

12

The benthos in the ecosystem

So far I have taken the benthos pretty much in isolation, dealing only in the previous chapter with the link between primary production and the input of organic matter to the sediment. In this final chapter I shall attempt to integrate the benthos into the complete ecosystem. Much of the foregoing has followed the traditional path of science, namely reductionism, where a problem is broken up into smaller, more manageable units. Recently there has been a change towards more holistic studies, where the behaviour of the whole system is studied without attempting to understand the functioning of all the parts. This is usually done by building models.

But what is a model and what can we expect from model building? A model is a simplification of a system and, therefore, can only represent a limited view of the behaviour of the whole system. Yet a model should have the important functional attributes of the real system, and it is in achieving a balance between realism and abstraction that the skill of the modeller lies.

The first stages in building a model involve putting the conceptualised system into a diagrammatic form. Usually this takes the form of a block diagram, where the units of the system are placed in compartments connected by arrows representing, for example, the flow of energy. The content of a compartment is called a variable and the input to the compartment a forcing function.

Quite often one has accumulated a wealth of data on the environmental variables in a given area and on the changes in animal numbers and biomass through time. By looking at correlations between variables and abundances it may be guessed that a certain relationship exists between the two, so a simple mathematical model is erected and tested on new data. Thus, one aim of a model is to formalise the relationship between field measurements and to test whether assumptions are true or not. If more is known about the relationships between a number of parts of the system, then these parts can be combined into a more complex model. Running this in a computer may reveal, properties called emergent properties, that were not obvious from treating each entity on its own. Therefore the second important aspect of modelling is that it can generate hypotheses.

If the model produced is reasonably realistic and gives good agreement with the field data, then it may be possible to use it to predict future states of the system. Often it would be quite impossible in a field situation to vary all the variables that one would wish to, but on a computer it is often a simple matter to simulate complex environmental changes. Thus, prediction is a third possibility with an ecosystem model.

In modelling one is interested not so much in describing the relationships between variables, but more in the mechanisms involved. For example it is relatively easy to establish that temperature may only be an indirect measure of the true mechanistic relationship between the process and the environment. In order to illustrate the modelling approach I will use two examples: Steele's model of the North Sea and the Baltic Sea model by Jansson.

12.1 The North Sea model

The North Sea model is concerned largely with plankton dynamics, but it does have a benthic component. The first step was to erect a compartmental model and then attempt to fit energy flow to the various compartments (Fig. 12.1).

Steele (1975) suggests that around 90 g C m^{-2} is a reasonable estimate of the particulate organic matter available from primary production, which is equivalent to 3768 kJ m^{-2} yr^{-1} (as 1 g carbon is equivalent to 41.868 kJ). The zooplankton eat all the phytoplankton, excreting about 30% as faecal material (1256 kJ m^{-2} yr^{-1}), which falls to the sea-bed, and producing 732 kJ m^{-2}, which is 19% of the primary production and is a rather high figure for efficiency. Other data indicate, though, that in marine systems efficiency does approach 20% with invertebrates and is not as low as the 10% assumed for terrestrial systems.

Half the herbivores are presumed eaten by invertebrate predators, such as ctenophores, medusae and jellyfish, the other half by pelagic fish. Fish standing stock is fairly well known, and the average number of pelagic fish in the North Sea is estimated at 4×10^6. This gives a total yield for pelagic fish of 8.0 g wet wt m^{-2} or 33.5 kJ m^{-2}, which represents a conversion efficiency from the herbivore trophic level of 10%. This value is reasonable and suggests that the estimate for energy input from the herbivores is of the right order.

The 1256 kJ m^{-2} that settles out as organic matter from zooplankton faeces has to account for, among other things, a total demersal fish production of 1.3×10^6 tonnes, which is equivalent to 2.6 g wet weight m^{-2} or 11 kJ m^{-2}. It is assumed that the organic material has first to be broken down into utilisable material. Bacteria have an efficiency of, at most, 30% and thus if all the material has to go

Fig. 12.1. Food web of The North Sea from Steele's model (1975). Figures within compartments are production in units of kJ m^{-2}; those between compartments are flows in kJ m^{-2} yr^{-1}. Broken lines indicate uncertainty.

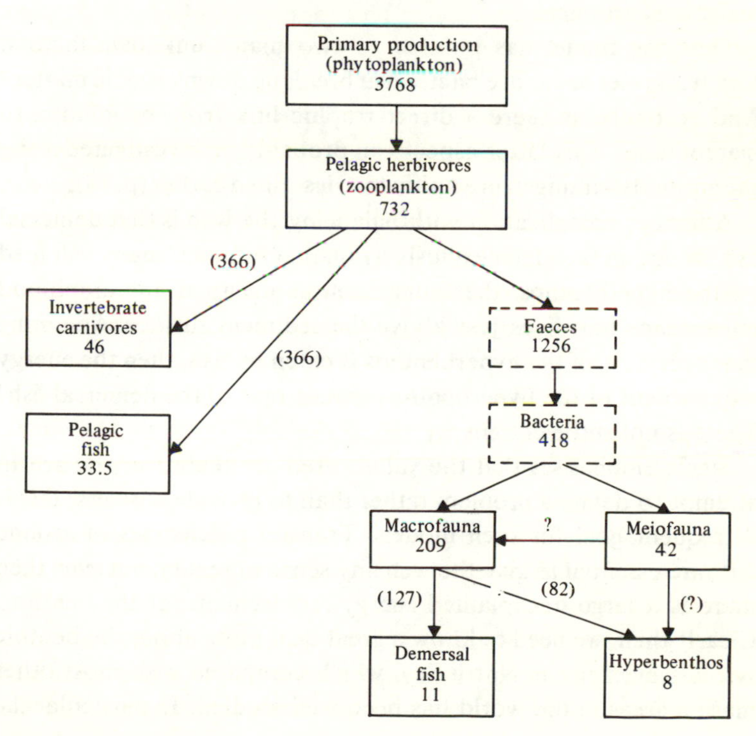

through bacteria before being used by the benthos then only 418 kJ m^{-2} of the original 1256 kJ m^{-2} will be available.

The standing stock of macrobenthos in the North Sea ranges from 0.6 to 1.6 g m^{-2}, and its production would require, say, 209 kJ m^{-2}. But for 209 kJ m^{-2} of macrobenthic production to have come from the 418 kJ m^{-2} of bacterial production necessitates a conversion efficiency of 50%, which is way above the known efficiencies for the macrofauna. Furthermore, the meiofauna has yet to be considered.

The meiofaunal standing crop is around 8 kJ m^{-2}, but the production figure is of course much higher than that say, by a factor of around 5, and can be estimated as 42 kJ m^{-2}. But we have already assumed that the macrofauna utilise all the bacterial production at an efficiency of around 50%, so there simply is not sufficient energy to sustain the meiofauna. The macrofauna could of course obtain some energy direct from the meiofauna, but the data on macrofauna–meiofauna predation are equivocal and it is not possible to put realistic figures here.

Thus the model has pointed to two major unknown factors. Firstly, how efficient are bacteria at breaking down organic matter? And secondly, is there a direct trophic link from meiofauna to macrofauna? This latter aspect can probably be investigated using the antibody–antingen precipitin test described earlier (p. 48).

A further complication with balancing the web is that demersal fish do not, in fact, feed exclusively on macrofauna; many fish feed on the hyperbenthos, the fauna such as decapod and amphipod crustaceans that lives just above the sediment surface. Assuming that only 10% of the hyperbenthos is eaten by fish, then the energy requirement of the hyperbenthos equals that of the demersal fish! There is not sufficient energy.

Steele emphasises that the values used are tentative, and are an attempt to define a problem rather than to provide answers; this is a frequent goal for such models. Transfer efficiencies of around 20% are essential to give the web any sense of reality, but even then there is a large unexplained energy requirement for the benthos. Clearly then, we need to know a great deal more about the benthic systems even in the North Sea, which compared with most other marine areas in the world has been well studied. In particular the efficiency and utilisation of organic matter by bacteria is largely guesswork, as is the amount of energy required by the hyperbenthos. Production data for meiofauna are lacking and the link between macrofauna and meiofauna is relatively unknown. The questions are legion.

Here, then, is an example of how a simple model has specified a number of key questions that need to be answered. The questions are part of an immensely important practical problem, that of setting rational fishing policies for the North Sea.

Steele's model was really concerned with the dynamics of the planktonic system, and used a series of simulations to suggest that the important control link in the plankton was that between the herbivores and primary production. This suggestion stimulated a flurry of research and arguments that the crucial link was, rather, between the carnivores and herbivores. The model that Steele used was a complex dynamic one and the example used here is merely a simple compartmentalised model. The data base was not good enough to do simulations of the benthic system.

A somewhat more complex benthic model has been erected for the fauna of the Baltic Sea by Jansson (1978) and his colleagues (Jansson & Wulf, 1977). The Baltic is a brackish water sea with a highly impoverished marine fauna. This fact was used to advantage by Jansson, for it allowed the model to be much simpler than would have been necessary for a typical marine environment containing many more species.

12.2 The Baltic Sea model

The Baltic Sea is the largest brackish-water area in the world; it covers 365 000 km^2 and has an average salinity of 6–7‰. The low salinity results in the species number inside the sill in the Øresund off Copenhagen being reduced by 90%. The low salinity is maintained by freshwater outflows from rivers, but there are inflows of high-salinity North Sea water in the form of a continuous flow of bottom water and a large periodic pulse in the autumn and winter. There are no tides in the Baltic so the salinity regime is stable and the water stratified; this results in a residence time for the water of 25–40 years. The major ecological compartments are the primary producers in the water column above the stratified layer and the phytal system covering the rocky archipelagos, and the consumers with the soft bottoms. There is a marked seasonal cycle which begins in December–January when the water is less stratified and inorganic salts are carried to the upper water layers. When the ice melts in spring, high insolation causes a marked phytoplankton bloom, which may result in large amounts of the blue-green algo *Nodularia spumigena*. The bloom is typically followed by a bloom of zooplankton. The nutrients in the surface layers also stimulate the algal belt of the rocky shores, and both *Cladophora* and *Fucus* grow rapidly and serve as nursery-grounds for many species. The phytoplankton provide a large input of settling organic matter; once on the bottom it is first broken down by ciliates prior to final breakdown by bacteria. In the autumn, storms remove large amounts of *Cladophora* and *Fucus* which sediment down to the bottom and are carried away in currents.

The group studying the Baltic was particularly concerned with one notable problem, that of the increased input of organic material from rivers and sewage that is resulting from the industrialisation and population growth in the countries bordering the Baltic. The largely stagnant waters of the deep basins in the Baltic cannot cope with the extra organic matter, and as a result all the oxygen is being used up and areas created where there is no oxygen at all, only hydrogen sulphide. Data collected over many decades suggest that the problem has been getting worse and that the periodic flushing by North Sea water no longer fully restores the basins. Over one-quarter of the Baltic bottom (100 000 km^2) is anoxic and devoid of life. Clearly, here is a practical problem of immense importance; how much extra organic matter can the Baltic ecosystem take without becoming completely dead? If the processes controlling the breakdown of the organic matter were more fully understood, more careful planning options could be taken.

The first step was to build a model incorporating known data, to establish, as Steele had done, what crucial data were missing. The Swedish group under Bengt-Owe Jansson elected to use the energy-circuit models pioneered by Tom Odum of the University of Florida.

Symbols were used to represent the processes concerned, because Odum believed that it would be too complex to represent each step by a series of differential equations. The symbols themselves stem from a consideration of basic thermodynamics. Three fundamental points must be understood: Firstly that the system must obey the first law of thermodynamics, which says that energy can be transformed but not destroyed; secondly that feedback and other interactions are vital to the functioning of the system; and thirdly that the language is unified by the use of basic physical and biological principles. In general the symbols are an extension of the compartmental boxes used in the previous models, but have the added advantage that they are easily translatable into a computer language.

Fig. 12.2 shows a compartmental diagram of the main processes involved. A source of energy, in this case the sun, can also be regarded as a forcing function, since the sun 'forces' photosynthesis in plants. A heat sink is a byproduct of work done, such as maintenance metabolism, and represents the energy degraded into heat during

Fig. 12.2. Flow of energy and matter within a coastal area of the Baltic Sea, depicted using energy-circuit language. A, primary production; B, secondary production; C, organic matter; D, sedimentation; E, sedimented matter; F, bottom fauna; G, nutrients. (After Jansson, 1978.)

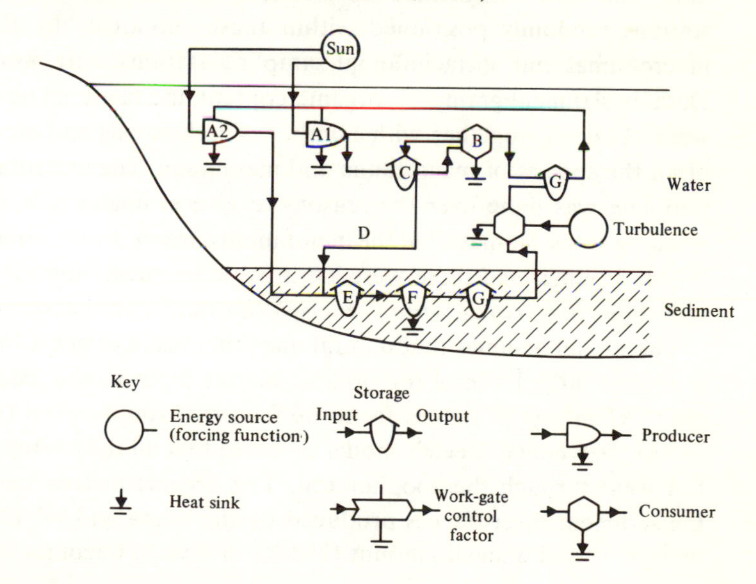

the process. A storage variable represents a situation where no energy is generated in moving energy or matter into and out of the storage (for example the flow of nitrate into the sea). A work-gate symbol is used for a flow of energy or matter that is acted upon by another factor (for example, the action of limiting nitrogen on plant production). The consumer symbol is a combination of a storage and work-gate. Here energy can be stored against a gradient and used in self-maintenance (for example, by an organism or group of organisms which use energy in order to obtain and metabolise food). Finally, the producer symbol represents green plants, which capture energy from the sun and use it to produce energy-rich reduced carbon. But as this energy is reoxidised during respiration, symbol used for a producer is that of a cycling-receptor. The symbols are then connected by flow pathways. In reality the complete Baltic model is built up from a series of integrated submodels, for the pelagial system, the algal belts, the hard bottom and the soft bottom. I shall discuss only the model for the soft bottom.

At the start of the project a diagrammatic compartmental model was built in order that factors on which there were no quantitative data could be clearly seen. Much effort was given to obtaining data on the quantity of macrofauna and meiofauna over an area of 128 km^2. The area was divided up into five sub-areas and sampling stations randomly positioned within these sub-areas. In all, 38 macrofaunal and 36 meiofaunal sampling stations were worked. Data on sediment grain size, organic content and bacterial number were obtained, together with number of individuals and biomass of all the species of macrofauna and meiofauna. The macrofaunal sampling was done over the seasons to give estimates of average standing stock biomass, but the meiofaunal data were only obtained for the summer and these were assumed to be mean biomass. Fig. 12.3 shows Sven Ankar's rough energy-flow model for the total area.

The model assumes that natural mortality (excluding predation) is around 30%. Little of the organic matter input to the sediment (2423 kJ m^{-2} yr^{-1}) comes from the faeces of zooplankton (91 kJ m^{-2} yr^{-1}), whereas Steele's model assumed that all the phytoplankton went through the zooplankton. The organic matter reaching the sediment meets faeces produced by the fauna and fish (388 kJ m^{-2} yr^{-1}) and a small amount (212 kJ m^{-2} yr^{-1}) becomes buried

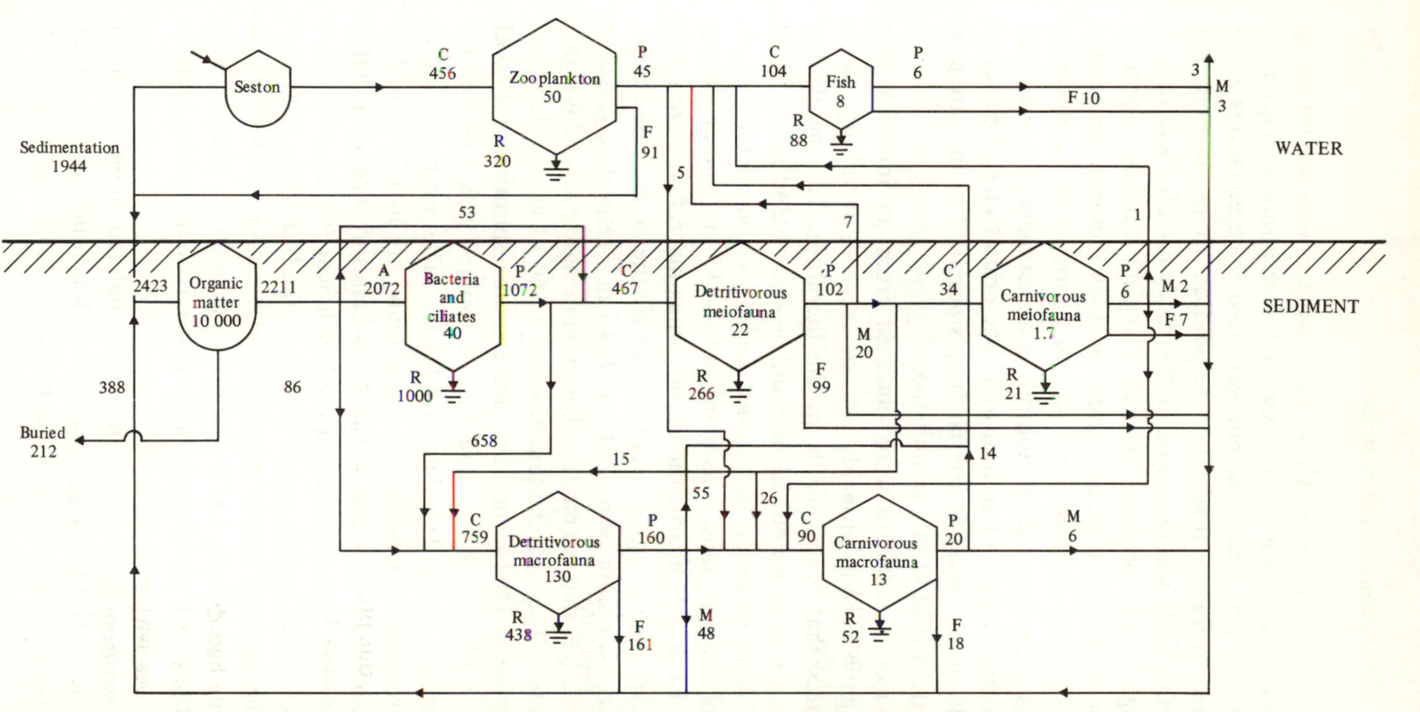

Fig. 12.3. Speculative energy-circuit model of the soft-bottom ecosystem of Askö-Landsort, Baltic Sea. Flows are in kJ m^{-2} yr^{-1} and storages in kJ m^{-2}. A, assimilation; C, consumption; F, faeces; R, respiration; M, natural mortality (fixed at 30%). Seston is phytoplankton and particulate organic matter. Other figures show predator–prey links except where stated. (Modified from Ankar, 1977.)

and no longer available. All of the consumers have energy going to a heat sink; this is the energy used in respiration; respiration was not, in fact, measured, but estimated from equations in the published literature that relate respiration to production. But production itself was only measured directly for a few species and most figures were obtained by multiplying the number of generations per year by the average standing stock biomass. So the production figures are approximations of approximations. Bacterial biomass was calculated by multiplying an estimate of numbers by the assumed weight of an individual bacterium. The efficiency of bacteria in this model is around 50%, since 1072 kJ m^{-2} are produced from 2082 kJ m^{-2} assimilated.

The meiofaunal data were obtained from an intensive and thorough study and probably provide the best overall figures on efficiences that are available in the literature. The meiofaunal detritivores consume 467 kJ m^{-2} and produce 102 kJ m^{-2}, with 99 kJ m^{-2} going to faeces. By contrast the carnivorous meiofauna are relatively unimportant, consuming only 34 kJ m^{-2} and producing only 6 kJ m^{-2}. Similarly, the detritivorous macrofauna are much more important energy consumers than are the macrofaunal carnivores, consumption being 759 and 90 kJ m^{-2} respectively. Ankar acknowledges that the model is based on meagre data and is highly speculative. The total benthic biomass is 260 kJ m^{-2}, with a total production or 1400 kJ m^{-2} yr^{-1} and a total respiration of 2100 kJ m^{-2} yr^{-1}. Most of this production is consumed within the benthic system itself and little is left over for fish. The production ratios of micro-organisms:meiofauna:macrofauna are 10:1:2. Table 12.1 summarizes a comparison of the Askö-Landsort data with that of Kiel Bay and Steele's model for the North Sea. Primary production is almost twice as high in the Baltic as in the North Sea, but otherwise the data are similar. Kiel Bay had overall much higher production, due probably to the shallow depth and the closer coupling of nutrients to primary production that is found with a continuously unstable water mass.

Steele had difficulty in balancing the high faunal biomass in the North Sea with the energy input from sedimenting organic matter. It is clear from Ankar's (1977) model of the Baltic that most of the organic matter is actually recycled within the benthic system. The rough speculations on energy flow indicate that the sedimenting organic matter is sufficient to balance the energy budget and that the Baltic is predominantly a detritivore system. The bacteria play such a dominant role in the energy balance of the soft-bottom system that much more effort should be given to their study. Similarly, the approximations used in this model need to be replaced by more accurate figures. The model is, however, a more comprehensive one than Steele's as far as the benthic components are concerned and points the way to future benthic models.

Both of the models discussed represent the dynamic aspects of the system by a static figure of kJ $m^{-2} yr^{-1}$. The utility of such models in pinpointing large gaps in our knowledge has been clearly demonstrated; as models, however, they represent only the very first step. The next step is to develop a mathematical expression of the system that can then be run on a computer so as to study the behaviour of the whole ecosystem. For example, in the Baltic model one may wonder what would happen if the oxygen concentration in the benthic system were reduced, all other variables remaining constant. Knowing values of the important forcing functions (which in this case are sedimentation (J_1) and the diffusion-turbulence of oxygen (J_2) and storage values, the experiment can be simulated on

Table 12.1. *Comparison of three energy-flow models (in kJ m^{-2} yr^{-1})*

	Baltic (Askö-Landsort)[a]	North Sea[b]	Kiel Bay[c]
Primary production	6000	3600	9200
Sedimentation	2400	1200	4900
Secondary production			
Micro-organisms	1000	?	?
Meiofauna	100	80	?
Macrofauna	200	200	440
Carnivores	20	8	20
Demersal fish	5	10	30

Data from: [a] Ankar (1977); [b] Steele (1975); [c] Arntz & Brunswig (1976).

a computer. Field experiments of this type would of course, be enormously costly and time-consuming.

The energy-circuit language is ideally suited to direct transfer to an analogue computer. An analogue computer is a simple system where differential equations are solved electronically by connecting together resistors, capacitors, etc. The models that can be simulated, therefore, must be composed of series of differential equations. The various components within the computer can be summated, integrated and multiplied, and there is only one independent variable, time, and one dependent variable, voltage. Integration, for example, is done by using a capacitor that adds and subtracts electrical flows. The linking up of the components is called patching and is analogous to the programming of a digital computer.

The details of the links and the equations representing the links are not important here and therefore Fig. 12.4 shows only a simplified model of the energy flow and storage. Q_1 is the oxygen concentration, Q_2 the detritus and Q_3 the biomass of benthic animals. The equations representing the transfers of energy were substituted into an analogue diagram and, after scaling (so that the full scale of the potentiometer equals the maximal observed value), the simulation was done. When the oxygen supply is reduced detritus accumulates, since the animal consumption decreases and the population reaches a steady state at a lower biomass than at the start.

This is a very simple simulation and merely indicates the sorts of results that can be achieved. From running a broader simulation of the Baltic model, Jansson and Wulf (1977) found that the factor really critical to the behaviour of the system was the exchange processes between the Baltic and the Kattegat. Thus again, the areas where more information are needed were pinpointed.

The above simulations were done on an analogue computer, which, in fact, very few scientists use today. Analogue machines are used with continuous data whereas digital computers process discrete numerical information. Since most data are discrete, and the analyses that can be done on a digital machine are not limited to differential equations, digital computers are preferred and are widely used. They do, however, require that special languages, such as Fortran, are used to convey information to the machine. Explanations of Fortran programs used in simulation studies of benthic systems are beyond the scope of this book. Fortran is the most widely used of present-day languages, but is over 20 years old and increasingly people are moving over to the more modern, simpler and easier to use languages, such as SIMULA. With digital machines, exceedingly complex simulations can be done.

The compartmental diagram models of the North Sea and Baltic merely used summated values of influxes of organic matter over a year, or production and respiration calculated over a year, and did

Fig. 12.4. Simulation on an analogue computer of the benthos of the Baltic Sea under normal conditions and with reduced oxygen. (Simplified from Janson & Wulff, 1977.)

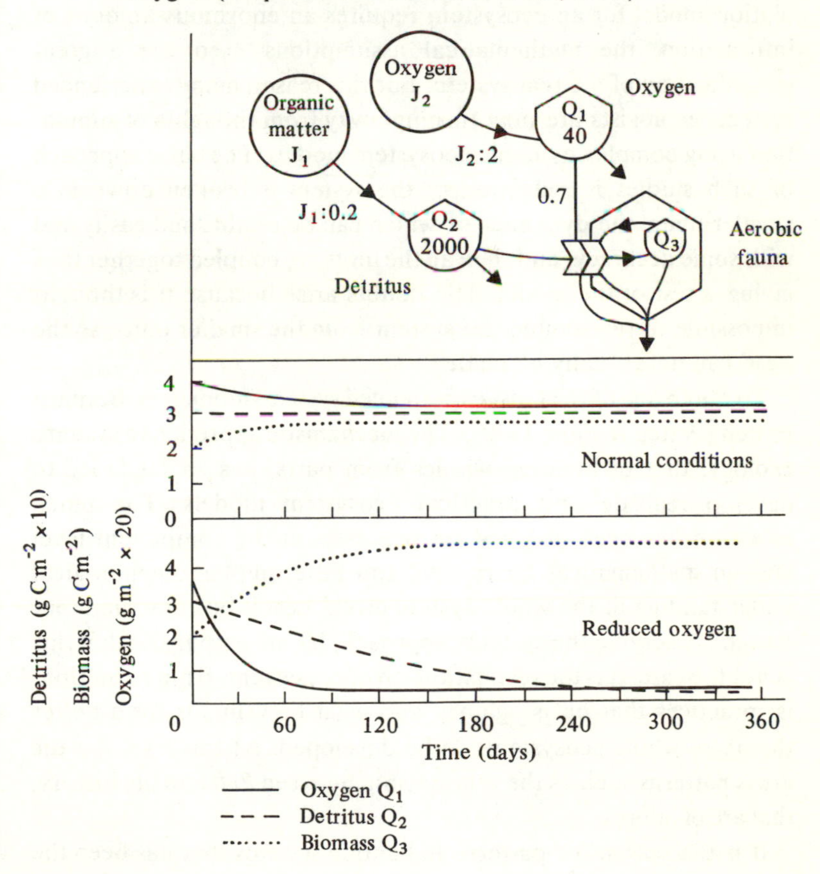

not simulate the variability from day to day or even month to month. The simulation in Fig. 12.4 gives an example of a change in one variable, oxygen, and the consequences this has for the benthos; it is based on simple equations derived from field observations. If the whole benthic system for the Baltic (Fig. 12.2) is to be simulated then each link in the chain must be represented by an equation. An analysis of this type (flow analysis) has been done on an Antarctic marine ecosystem (Patten & Finn, 1979). The whole set of equations, together gives a complex hypotheses of how the system functions, but each equation is a subhypothesis which can be independently tested and modified. But for ease of computation one often has to assume a linear relationship between variables when most relationships in nature are actually non-linear. Thus while a complex simulation model for an ecosystem requires an enormous amount of information, the mathematical assumptions used are a great simplification of the real system. For this reason many experienced systems ecologists are now turning away from thoughts of simulation using complex dynamic ecosystem models. The basic approach of such studies is reductionist: the system is broken down into smaller units, the dynamics of which can be established easily and with some accuracy, and then all the units are coupled together thus giving a test of the model. The doubts arise because it is thought impossible to reassemble the system from the smaller units, so the 'test' can never really be made.

To quote one of the most experienced systems modellers, Bernard Patten (Patten & Finn, 1979), 'The mechanistic approach to systems ecology, that of building wholes from parts, has so far failed to develop realistic and practical ecosystem models. Too much attention has been lavished on processes at the component level and on mathematical finery, and too little on phenomenological understanding at the whole-system level.' Lest it be construed that Patten is decrying the systems approach, let me emphasise that this is not the case. It is the reconstitution of ecosystems from reductionist practices that he is against, and what he wants is for a better theory of whole ecosystems to be developed. At this level it is the gross patterns such as the relationship between $P/\bar{B}$ and life history, that are of interest.

It is the search for patterns in benthic systems that has been the main theme of this book, starting from the pattern of how individuals are distributed among species, through patterns of diversity to patterns of energy flow through a complete benthic community. And surely it is here where all branches of ecology meet, and marine ecology no longer stands alone as a rather backward and merely descriptive science but becomes integrated into the discipline as a whole.

REFERENCES

Anger, K. (1975). On the influence of sewage pollution on inshore benthic communities in the south of Kiel Bay. II. Quantitative studies on community structure. *Helgoländer Wissenschaftliche Meeresuntersuchungen*, **27**, 408–38.

Ankar, S. (1977). The soft bottom ecosystem of the Northern Baltic proper with special reference to the macrofauna. *Contributions from the Askö Laboratory, University of Stockholm, Sweden*, **19**, 1–62.

Arntz, W. E. & Brunswig, D. (1976). Studies on structure and dynamics of macrobenthos in the western Baltic carried out by the joint research programme 'Interaction sea–sea bottom'. In *Proceedings of the 10th European Symposium on Marine Biology* (Ostend), vol. 2, ed. G. Persoone & G. Jaspers, pp. 17–42. Wettern, Belgium: Universa Press.

Bayne, B. L. (1980). Physiological measurements of stress. In *Biological Effects of Marine Pollution and the Problems of Monitoring*, ed. A. D. McIntyre & J. B. Pearce. *Rapport et procès-verbaux des réunions Conseil permanent international pour l'exploration de la mer*, in press.

Bliss, C. I. (1966). An analysis of some insect trap records. In *Proceedings of an International Symposium on Classical and Contagious Distributions* (Montreal, 1963), ed. G. Patil, pp. 385–97. Calcutta: Statistical Publishing Co.

Bliss, C. I. (1970). *Statistics in Biology*, vol. 2. New York: McGraw-Hill. 639pp.

Boysen-Jensen, P. (1919). Valuation of the Limfjord: I. *Report of the Danish Biological Station*, **26**, 1–24.

Buchanan, J. B. (1971). Measurements of the physical and chemical environment. In *Methods for the Study of Marine Benthos*, ed. N. Holme & A. D. McIntyre, *IBP Handbook 16*, pp. 30–58. Oxford: Blackwell Scientific Publications.

Buchanan, J. B. & Longbottom, M. R. (1970). The determination of organic matter in marine muds: the effect of the presence of coal and the routine determination of protein. *Journal of Experimental Marine Biology and Ecology*, **5**, 158–69.

Buchanan, J. B., Sheader, M. & Kingston, P. R. (1978). Sources of variability in the benthic macrofauna off the south Northumberland coast, 1971–6. *Journal of the Marine Biological Association of the United Kingdom*, **58**, 191–210.

Burke, M. V. & Mann, K. H. (1974). Productivity and production: biomass ratios of bivalve and gastropod populations in an Eastern Canadian estuary. *Journal of the Fisheries Research Board of Canada*, **31**, 167–77.

Clifford, H. T. & Stephenson, W. (1975). *An Introduction to Numerical Classification*. New York: Academic Press. 224pp.

Crisp, D. J. (1971). Energy flow measurements. In *Methods for the Study of Marine Benthos*, ed. N. A. Holme & A. D. McIntyre, *IBP Handbook 16*, pp. 197–279. Oxford: Blackwell Scientific Publications.

Cushing, D. M. (1975). *Marine Ecology and Fisheries.* London: Cambridge University Press. 278pp.

Davies, J. M. (1975). Energy flow through the benthos in a Scottish sea loch. *Marine Biology*, **31**, 353–62.

Dayton, P. K. & Hessler, R. R. (1972). Role of biological disturbance in maintaining diversity in the deep sea. *Deep-Sea Research*, **19**, 199–208.

Dörjes, J. & Howard, J. D. (1975). Estuaries of the Georgia Coast, USA: sedimentology and biology. IV. Fluvial–marine transition indicators in an estuarine environment, Ogeechee River–Ossabow Sound. *Senckenbergiana maritima*, **7**, 137–9.

Eagle, R. A. & Hardiman, P. A. (1977). Some observations on the relative abundance of species in a benthic community. In *Biology of Benthic Organisms*, ed. B.F. Keegan, P. O'Ceidigh & P. J. S. Boaden, *11th European Symposium on Marine Biology* (Galway, 1976), pp. 197–208. Oxford: Pergamon Press.

Elliot, J. M. (1971). *Some Methods for the Statistical Analysis of Samples of Benthic Invertebrates. FBA Scientific Publication* 144pp. **25**. Ambleside: Freshwater Biological Association.

Elton, C. S. (1966). *The Pattern of Animal Communities.* London: Methuen. 181pp.

Feller, R. J., Taghon, G. L., Gallagher, E. D., Kenny, G. E. & Jumars, P. A. (1979). Immunological methods for food web analysis in a soft-bottom benthic community. *Marine Biology*, **54**, 61–74.

Fenchel, T. (1975). Character displacement and coexistence in mud snails (Hydrobiidae). *Oecologia* (Berlin), **20**, 19–32.

Fenchel, T. (1978). The ecology of micro- and meiobenthos. *Annual Review of Ecology and Systematics*, **9**, 99–121.

Fenchel, T. & Riedl, R. J. (1970). The sulphide system: a new biotic community underneath the oxidised layer of marine sand bottoms. *Marine Biology*, **7**, 255–68.

Ferguson-Wood, E. J. (1965). *Marine Microbial Ecology.* London: Chapman & Hall. 242pp.

Folk, R. L. (1968). *Petrology of Sedimentary Rocks.* Austin: Hemphills. 170pp.

Gerlach, S. A. (1971). On the importance of marine meiofauna for benthos communities. *Oecologia* (Berlin), **6**, 176–90.

Gerlach, S. A. (1972). Die Produktionsleistung des Benthos in der Helgoländer Bucht. *Verhandlungsbericht der Deutschen Zoologischen Gesellschaft*, **65**, 1–13.

Grassle, J. F. & Grassle, J. P. (1974). Opportunistic life-histories and genetic systems in marine benthic polychaetes. *Journal of Marine Research*, **32**, 253–84.

Grassle, J. F. & Sanders, H. L. (1973). Life-histories and the role of disturbance. *Deep-Sea Research*, **20**, 643–59.

Gray, J. S. (1965). The behaviour of *Protodrilus symbioticus* (Giard) in temperature gradients. *Journal of Animal Ecology*, **34**, 455–61.

Gray, J. S. (1966*a*). The response of *Protodrilus symbioticus* (Giard) to light. *Journal of Animal Ecology*, **35**, 55–64.

Gray, J. S. (1966*b*). Selection of sands by *Protodrilus symbioticus* (Giard). *Veröffentlichungen des Instituts für Meeresforschung in Bremerhaven*, **2**, 105–16.

Gray, J. S. (1966*c*). The attractive factor of intertidal sands to *Protodrilus symbioticus. Journal of the Marine Biological Association of the United Kingdom*, **46**, 627–45.

Gray, J. S. (1966*d*). Factors controlling the localizations of populations of *Protodrilus symbioticus* Giard. *Journal of Animal Ecology*, **35**, 435–42.

Gray, J. S. (1974). Animal–sediment relationships. In *Oceanography and Marine Biology: An Annual Review*, ed. H. Barnes, vol. 12, pp. 223–61. London: Allen & Unwin.

Gray, J. S. (1977). The stability of benthic ecosystems. *Helgoländer Wissenschaftliche Meeresuntersuchungen*, **30**, 427–44.

Gray, J. S. (1978). The structure of meiofauna communities. *Sarsia*, **64**, 265–72.

Gray, J. S. (1979*a*). Pollution-induced changes in populations. *Philosophical Transactions of the Royal Society of London Series B*, **286**, 545–61.

Gray, J. S. (1979*b*). The development of a monitoring programme for Norway's coastal marine fauna. *Ambio*, **8**, 176–9.

Gray, J. S. & Mirza, F. B. (1979). A possible method for the detection of pollution-induced disturbance on marine benthic communities. *Marine Pollution Bulletin*, **10**, 142–6.

Grime, J. P. (1979). *Plant Strategies and Vegetation Processes.* New York: Wiley. 222pp.

Hargrave, B. (1978). Benthic communities. In *Biological Oceanographic Processes*, 2nd edn, ed. T. A. Parsons, M. Takahishi & B. Hargrave, pp. 176–264. Oxford: Pergamon Press.

Harris, R. P. (1973). Feeding, growth, reproduction and nitrogen utilization by the harpacticoid copepod, *Tigriopus brevicornis*. *Journal of the Marine Biological Association of the United Kingdom*, **53**, 785–800.

Heip, C. (1976). The calculation of eliminated biomass. *Biologische Jahrbok Dodonaea*, **44**, 217–25.

Heip, C. & Decraemer, W. (1974). The diversity of nematode communities in the southern North Sea. *Journal of the Marine Biological Association of the United Kingdom*, **54**, 251–5.

Heip, C. & Herman, R. (1979). Production of *Nereis diversicolor* O.F. Müller (Polychaeta) in a shallow brackish-water pond. *Estuarine and Coastal Marine Science*, **8**, 297–305.

Hickman, M. & Round, F. E. (1970). Primary production and standing crop of epipsammic and epipelic algae. *British Psychological Journal*, **5**, 247–55.

Holme, N. & McIntyre, A. D. (eds.) (1971). *Methods for the Study of Marine Benthos. IBP Handbook 16.* Oxford: Blackwell Scientific Publications. 334pp.

Hughes R. G. & Thomas, M. L. (1971). The classification and ordination of shallow-water benthic samples from Prince Edward Island, Canada. *Journal of Experimental Marine Biology and Ecology*, **7**, 1–39.

Hughes, R. N. (1970). An energy budget for a tidal-flat population of the bivalve *Scrobicularia plana* (DaCosta). *Journal of Animal Ecology*, **39**, 357–81.

Hulings, N. C. & Gray, J. S. (1971). A manual for the study of meiofauna. *Smithsonian Contributions to Zoology*, **78**, 84pp.

Hurlbert, S. N. (1971). The non-concept of species diversity: a critique and alternative parameters. *Ecology*, **52**, 577–86.

Huston, M. (1979). A general hypothesis of species diversity. *American Naturalist*, **113**, 81–101.

Inman, D. I. (1949). Sorting of sediments in the light of fluid mechanics. *Journal of Sedimentary Petrology*, **7**, 3–17.

Jansson, B. O. (1978). The Baltic: a systems analysis of a semienclosed sea. In

Advances in Oceanography, ed. H. Charnock & G. Deacon, pp. 131–83. New York: Plenum Press.

Jansson, B. O. & Wulff, F. (1977). Ecosystem analysis of a shallow sound in the northern Baltic. A joint study by the Askö group. *Contributions from the Askö Laboratory, University of Stockholm, Sweden*, **18**, 1–160.

Jumars, P. A. (1975). Environmental grain and polychaete species diversity in a bathyal benthic community. *Marine Biology*, **30**, 253–66.

King, C. E. (1964). Relative abundance of species in MacArthur's model. *Ecology*, **45**, 716–27.

Kirkegaard, J. B. (1978). Production by polychaetes on the Dogger Bank in the North Sea. *Meddelelser fra Danmarks Fiskeri-og Havundersøgelser*, **7**, 497–509.

Lasker, R., Wells, J. B. J. & McIntyre, A. D. (1970). Growth, reproduction, respiration and carbon utilization of the sand-dwelling harpacticoid copepod, *Asellopsis intermedia. Journal of the Marine Biological Association of the United Kingdom*, **50**, 147–60.

Lassig, J. & Lahdes, E. (1980). Biological monitoring and effects studies in the Baltic Sea: a review of activities carried out in Finland. In *Biological Effects of Marine Pollution and the Problems of Monitoring*, ed. A. D. McIntyre & J. B. Pearce. *Rapport et procès-verbaux des réunions. Conseil permanent international pour l'exploration de la mer*, in press.

Levinton, J. (1972). Stability and trophic structure in deposit-feeding and suspension-feeding communities. *American Naturalist*, **106**, 472–86.

Levinton, J. D. & Lopez, G. R. (1977). A model of renewable resources and limitation of deposit-feeding benthic populations. *Oecologia* (Berlin), **31**, 177–90.

Lindeman, R. L. (1942). The trophic-dynamic aspect of ecology. *Ecology*, **23**, 394–418.

MacArthur, R. H. (1957). On the relative abundance of bird species. *Proceedings of the National Academy of Sciences, USA*, **43**, 293–5.

MacArthur, R. H. & Wilson, E. O. (1967). *The Theory of Island Biogeography*. Princeton: Princeton University Press. 203pp.

McIntyre, A. D. (1969). Ecology of marine meiobenthos. *Biological Reviews*, **44**, 245–90.

McIntyre, A. D. (1970). The range of biomass in intertidal sand, with special reference to the bivalve *Tellina tenius. Journal of the Marine Biological Association of the United Kingdom*, **50**, 561–76.

McIntyre A. D. & Pearce, J. B. (eds.) (1980). *Biological Effects of Marine Pollution and the Problems of Monitoring. Rapport et procès-verbaux des réunions. Conseil permanent international pour l'exploration de la mer*, in press.

Margalef, R. (1968). *Perspectives in Ecological Theory*. Chicago: University of Chicago Press, 111pp.

May, R. M. (1975). Patterns of species abundance and diversity. In *Ecology and Evolution of Communities*, ed. M. L. Cody & J. M. Diamond, pp. 81–120. Cambridge, Mass.: Belknap Press.

Mills, E. L. (1969). The community concept in marine zoology, with comments on continua and instability in some marine communities: a review. *Journal of the Fisheries Research Board of Canada*, **26**, 1415–28.

Moore, P. G. (1973). The kelp fauna of northeast Britain. II. Multivariate classification: turbidity as an ecological factor. *Journal of Experimental Marine Biology and Ecology*, **13**, 127–63.

Paine, R. T. (1966). Food web complexity and species diversity. *American Naturalist*, **100**, 65–75.

Pamatmat, M. M. (1968). Ecology and metabolism of a benthic community on an intertidal sandflat. *Internationale Revue der gesamten Hydrobiologie*, **53**, 211–98.

Pamatmat, M. M. & Banse, K. (1969). Oxygen consumption by the seabed. II. *In situ* measurements to a depth of 180 m. *Limnology and Oceanography*, **14**, 250–9.

Patten, B. C. & Finn, J. T. (1979). Systems approach to continental shelf ecosystems. In *Theoretical Systems Ecology*, ed. E. Halfon, pp. 183–212. New York: Academic Press.

Pearson, R. H. (1975). The benthic ecology of Loch Linnhe and Loch Eil, a sea-loch system on the west coast of Scotland. IV. Changes in the benthic fauna attributable to organic enrichment. *Journal of Experimental Marine Biology and Ecology*, **20**, 1–41.

Pearson, T. H. & Rosenberg, R. (1978). Macrobenthic succession in relation to organic enrichment and pollution of the marine environment. *Oceanography and Marine Biology: An Annual Review*, vol. 16, pp. 229–311. Aberdeen: Aberdeen University Press.

Petersen, C. G. J. (1914). Valuation of the sea. II. The animal communities of the sea bottom and their importance for marine zoogeography. *Reports of the Danish Biological Station*, **21**, 44pp.

Petersen, C. G. J. (1915). On the animal communities of the sea bottom in the Skaggerak, the Christiania Fjord and Danish waters. *Reports of the Danish Biological Station*, **23**, 3–28.

Petersen, C. G. J. (1918). The sea bottom and its production of fish food. A survey of work done in connection with the valuation of the Danish waters from 1883–1917. *Reports of the Danish Biological Station*, **25**, 1–62.

Petersen, C. G. J. (1924). A brief survey of the animal communities in Danish waters. *American Journal of Science*, **7**, 343–54.

Peterson, C. H. (1980). Predation, competitive exclusion and diversity in the soft-sediment benthic communities of estuaries and lagoons. In *Ecological Processes in Coastal and Marine Systems*, ed. R. J. Livingstone, pp. 233–64. New York: Plenum Press.

Pielou, E. C. (1975). *Ecological Diversity*. New York: Wiley.

Preston, F. W. (1948). The commonness and rarity of species. *Ecology*, **29**, 254–83.

Rachor, E. & Gerlach, S. A. (1978). Changes in a sublittoral sand area of the German Bight, 1967 to 1975. *Rapport et procès-verbaux des réunions. Conseil permanent international pour l'exploration de la mer*, **172**, 418–31.

Reise, K. (1977). Predator exclusion experiments in an intertidal mud flat. *Helgoländer wissenschaftliche Meeresuntersuchungen*, **30**, 263–71.

Reish, D. J. (1959). A discussion of the importance of screen size in washing quantitative marine bottom samples. *Ecology*, **40**, 307–9.

Rhoads, D. C. (1974). Organism–sediment relations on the muddy sea-floor. In *Oceanography and Marine Biology: an Annual Review*, ed. H. Barnes, vol. 12, pp. 263–300. London: Allen & Unwin.

Rhoads, D. C. & Young, D. K. (1970). The influence of deposit-feeding organisms on sediment stability and community trophic structure. *Journal of Marine Research*, **28**, 150–78.

Robertson, A. I. (1979). The relationship between annual production: biomass ratios and lifespans for marine macrobenthos. *Oecologia* (Berlin), **38**, 193–202.

Sanders, H. L. (1956). Oceanography of Long Island Sound, 1952–4. X. The biology of marine bottom communities. *Bulletin of the Bingham Oceanographic Collection*, **15**, 345–414.

Sanders, H. L. (1958). Benthic studies in Buzzard's Bay. I. Animal–sediment relationships. *Limnology and Oceanography*, **3**, 245–58.

Sanders, H. L. (1968). Marine benthic diversity: a comparative study. *American Naturalist*, **102**, 243–82.

Sanders, H. L., Hessler, R. R. & Hampson, G. R. (1965). An introduction to the study of the deep-sea benthic faunal assemblages along the Gay Head–Bermuda Transect. *Deep-Sea Research*, **12**, 845–67.

Smith, K. L., Burns, H. A. & Teal, J. M. (1972). *In situ* respiration of benthic communities in Castle Harbour, Bermuda. *Marine Biology*, **12**, 196–9.

Steele, J. H. (1975). *The Structure of Marine Ecosystems*. Cambridge, Mass: Harvard University Press. 128pp.

Stenseth, N. C. (1979). Where have all the species gone? On the nature of extinction and the Red Queen hypothesis. *Oikos*, **33**, 196–227.

Stephenson, W. S. (1978). Analyses of periodicity on macrobenthos using constructed and real data. *Australian Journal of Ecology*, **3**, 321–36.

Stephenson, W., Williams, W. T. & Cook, S. G. (1971). Computer analyses of Petersen's original data on bottom communities. *Ecological Monographs*, **42**, 387–415.

Swedmark, B. (1964). The interstitial fauna of marine sand. *Biological Reviews*, **39**, 1–42.

Teal, J. M. & Kanwisher, J. (1961). Gas exchange in a Georgia salt marsh. *Limnology and Oceanography*, **6**, 388–99.

Thorson, G. (1957). Bottom communities (sublittoral or shallowshelf). In *Treatise on Marine Ecology and Palaeoecology*, vol. 1, *Ecology*, ed. J. W. Hedgpeth. *Memoirs of the Geological Society of America*, **67**, 461–534.

Vandermeer, J. H. (1972). Niche theory. *Annual Review of Ecology and Systematics*, **3**, 107–32.

Wieser, W. & Zech, M. (1976). Dehydrogenases as tools in the study of marine sediments. *Marine Biology*, **36**, 113–22.

Williams, C. B. (1964). *Patterns in the Balance of Nature and Related Problems in Quantitative Ecology*. New York: Academic Press. 324pp.

Woodin, S. A. (1974). Polychaete abundance patterns in a marine soft-sediment environment. The importance of biological interactions. *Ecological Monographs*, **44**, 171–87.

Woodin, S. A. (1976). Adult–larval interactions in dense faunal assemblages: patterns of abundance. *Journal of Marine Research*, **34**, 25–41.

Young, D. K. & Rhoads, D. C. (1971). Animal–sediment relations in Cape Cod Bay, Massachusetts. I. A. transect study. *Marine Biology*, **11**, 242–54.

INDEX

References in **bold** type refer to chapters; references in *italic* type refer to tables and figures.